T0182090

In Search of the Next Memory

Roberto Gastaldi · Giovanni Campardo
Editors

In Search of the Next Memory

Inside the Circuitry from the Oldest
to the Emerging Non-Volatile Memories

 Springer

Editors
Roberto Gastaldi
Redcat Devices s.r.l
Milan
Italy

Giovanni Campardo
STMicroelectronics
Milan
Italy

ISBN 978-3-319-83806-9   ISBN 978-3-319-47724-4   (eBook)
DOI 10.1007/978-3-319-47724-4

Front Cover: Giovanni Campardo *The Lily Water* (Le ninfee), Oil on canvas, 50 x 70, 2012 [Francesca Carsana, private collection] [Photographed by Alice Campardo]
The lotus flower meaning is different between cultures, though in fact they share many similarities: purity, spiritual awakening and faithfulness, spiritual enlightenment, sun and creation, rebirth, victory of the spirit over that of wisdom, intelligence and knowledge. Especially the red lotus is related to the heart, and the Lotus flower meaning is associated with that of love and compassion.

Printed on acid-free paper

This Springer imprint is published by Springer Nature
The registered company is Springer International Publishing AG
The registered company address is: Gewerbestrasse 11, 6330 Cham, Switzerland

Preface

Over the last few years, new technologies have dramatically taken hold of the non-volatile memories industry, which had been dominated for a long time by Flash technology.

The reason can be found partly in the fact that it has become increasingly difficult to overcome Flash's technological and physical limits in order to continue following the path of dimensional reduction.

Even if we know from experience that those limits are, in fact, partial and that they keep moving yet forward as technology progresses, it is evident that more and more financial resources and efforts are needed to finally overcome them. However, an incentive for change also comes from new application requirements that demand real-time elaboration and permanent memorization of a large amount of data.

The current memory system, based on fast and temporary memory (DRAM) and permanent memory (NAND), seems to be insufficient to meet the new demands. In addition, the development of DRAM technology is facing difficulties that are similar to those found with NAND.

For all these reasons, many expectations have been put on new technologies in the hope of finding one of them to be the ideal memory.

Since there are various new proposals based on very different physical principles, it is not always easy to clearly identify the pros and cons of each approach and to recognize analogies and differences. The idea behind this book is to provide a little help in this direction.

We start with a brief description of mainstream Flash technology, the main problems that make it difficult to proceed further with it, and the solutions that have been adopted to assure its survival. Then, we turn to breakthrough approaches, trying to compare some proven emerging technologies while analyzing their operating principles and the essential building blocks of related device architecture.

We believe that a fundamental theme is programming algorithms, and we pay careful attention to them. They are very important in the current NAND technology. Despite the initial hopes to do without them, they are still crucial for some new emerging technologies. The same applies to on-board systems for error correction, which are now becoming an integral part of the new architecture for high-density memories.

Lastly, we briefly touch upon the theme of device sensitivity to high-energy radiation. Usually, it is an area of expertise for specialists and may seem an odd topic, considering the main theme of this book.

However, we wanted to add this part because the technology of the new memories has been considered very interesting also in this respect, raising the hope of decisive performance improvement for devices operating in a high-energy radiation environment.

Certainly, we do not expect this book to be exhaustive: such a dynamic field as non-volatile memories has generated and still generates many innovative ideas, some of them introducing new concepts, and others recalling already-known notions.

We decided to examine only those technologies that have developed to a certain level of maturity. Other technologies are still in an early stage of research, and we decided not to talk about them in this book, but this choice, however, should not be seen as a negative opinion about the possibility that some of them will be successful in the future.

This book could never have been written without the help and support of the many people who collaborated for its realization with enthusiasm and knowledge. Therefore, we would like to thank:

Roberto Bez, who wrote the introduction and with whom we shared many years of work in the central R&D group of STMicroelectronics in Agrate;

Agostino Pirovano, curator of the chapter dedicated to emerging memories' technological aspects;

Guido Torelli and Alessandro Cabrini for their valuable expertise in defining programming algorithms for PCM memories;

Andrea Fantini, who worked on programming ReRAM memories;

Paolo Amato and Marco Sforzin, who dedicated themselves to the error-correction system, a crucial aspect of new technologies;

Carmen Wolf and Christoph Baumann from the Springer publishing house, who helped us with competence in every step of the realization of this book and Ken Quinn for the meticulous linguistic revision process.

We would like to offer special thanks to our families who willingly accepted for fitting part of the time that otherwise would have been dedicated to them so that we could complete this project.

Our sincere hope is that this simple work can guide those who are interested in the world of emerging non-volatile memories, mainly from a designer prospective, but also from the user's point of view, and help those who wish to keep up with the new developments in memories that are achieving an increasingly crucial role in data processing systems of the present and very likely in the future.

A teacher once asked his four students to define what memory is.

After a few days, he met them again and asked if they had come up with the definition.

The first student said: "memory is the ability to keep and reproduce previous thoughts without the reasoning that gave rise to them being present, in the same way as reminiscence is the ability to recollect in our mind things we have learned." You are really good, the teacher said. You will go a long way with your philosophical skills.

The second student said: "memory is the ability to store information, from the simple details of everyday life to complex concepts such as knowledge of abstract geography or algebra. Surely it is one of the most extraordinary aspects of human behavior." Good, the teacher said. You will definitely become a great medical researcher.

The third student answered: "memory is part of a system that registers and stores data and instructions for further elaboration." I see you are analyzing technical aspects of memory, the teacher said. You will be a great technician.

The fourth student said: "memory enables us to preserve experiences, therefore to recollect anything in our heart that might invoke mercy, so that our heart can be touched by other People's. "The teacher after hearing this said: "I am your student."

Milan, Italy   Roberto Gastaldi
September 2016   Giovanni Campardo

Contents

Editors and Contributors

About the Editors

Roberto Gastaldi received the MS degree in electronic engineering from the Politecnico of Milano, Italy, in 1977 and in the same year he joined the Central R&D department of SGS-ATES (later STMicroelectronics) as a device engineer. In 1981, he joined the non-volatile memory department starting to design EPROMS and first EEPROMS. From 1993 to 2000, he directed Eprom design development group. In 1989, he started also to work on Flash-NOR technology and products development as a member of technical staff. From 2000 to 2005, he worked on SRAM and DRAM design and breakthrough technologies evaluation, developing PCM memory architecture and circuit design. From 2008 to 2010, he served as a manager of advanced design team at Numonyx, and from 2010 to 2014 he was with Micron Semiconductor Italy in the emerging memory department of central R&D where he worked on TRAM and STT-MRAM architecture development and explored novel sensing techniques and smart error correction engines for storage class memory applications.

He retired in January 2015. Presently, he works in the area of memory design and technology and radiation-hardening by design.

Mr. Gastaldi is co-author of papers and conference contributions on topics related to NVM design and holds more than 50 US-granted patents on memory design. He has been a member of the winner team of the "Innovator of the Year" award for EDN-2009 and a lecturer on selected topics on Microelectronics at Pavia University (Italy).

Mr. Gastaldi served as a member of ISSCC technical program memory subcommittee from 2009 to 2011 and co-chaired ISSCC2010 Non-volatile memory session in S. Francisco, February 2010.

In 2014, he was a member of CATRENE technical steering group for applications.

He is a member of IEEE.

Giovanni Campardo was born in Bergamo, Italy, in 1958. He received the laurea degree in nuclear engineering from the Politecnico of Milan in 1984. In 1997, he graduated in physics from the Universita' Statale di Milano, Milan.

After a short experience in the field of laser in 1984, he joined in the VLSI division of SGS (now STMicroelectronics) Milan, where, as a project leader, he designed the family of EPROM nMOS devices (512k, 256k, 128k, and 64k) and a look-up-table-based EPROM FIR in CMOS technology. From 1988 to 1992, after resigning from STMicroelectronics, he worked as an ASIC designer, realizing four devices. In 1992, he joined STMicroelectronics again, concentrating on Flash memory design for the microcontroller division, as a project leader. Here, he has realized a Flash + SRAM memory device for automotive applications and two embedded Flash memories (256k and 1M) for ST10 microcontroller family. Since 1994, he has been responsible for Flash memory design inside the Memory Division of SGS-Thomson Microelectronics where he has realized two double-supply Flash Memories (2M and 4M) and the single-supply 8M at 1.8 V. He was the design manager for the 64M multilevel Flash project. Up to the end of 2001, he was the product development manager for the Mass Storage Flash Devices in STMicroelectronics Flash Division realizing the 128M multilevel Flash and a test pattern to store more than 2 bit/cell. From 2002 to 2007, inside the ST Wireless Flash Division, he had the responsibility of building up a team to develop 3D Integration in the direction of System-in-Package solutions. From 2007 to 2010, he was the director of the Card Business Unit inside the Numonyx DATA NAND Flash Group. In 2010, Micron acquired Numonyx and, after a short time in Micron he moved in a company, as design director, for the EWS testing board design. In 2013, he joined Micron again as managed memory systems manager. From September 2014, he is in STM, in the Automotive Product Group, in the R&D design group.

He is author/co-author of more than 100 patents and some publications and author/co-author of the books: "Flash Memories," Kluwer Academic Publishers, 1999, and the book "Floating Gate Devices: Operation and Compact Modeling," Kluwer Academic Publishers, January 2004. Author of the book "*Design of Non-Volatile Memory*," Franco Angeli, 2000, and "*VLSI-Design of Non-Volatile Memories*," Springer Series in ADVANCED MICROELECTRONICS, 2005. "Memorie in Sistemi Wireless," Franco Angeli Editore, collana scientifica, serie di Informatica, 2005. "*Memories in Wireless Systems*," in CIRCUITS AND SYSTEMS, Springer Verlag, 2008. G. Campardo, F. Tiziani, M. Iaculo "*Memory Mass Storage*," Springer Verlag, March 2011. He was the co-chair for the "System-In-Package-Technologies" Panel discussion for the IEEE 2003 Non-Volatile Semiconductor Memory Workshop, 19th IEEE NVSMW, Monterey; Ca. Mr. Campardo was the co-guest editor for the Proceeding of the IEEE, April 2003, Special issue on Flash Memory and the co-guest editor for the Proceeding of the IEEE, Special issue on 3D Integration Technology, January 2009. He was lecturer in the "Electronic Lab" course at the University Statale of Milan from 1996 to 1998. In 2003–2005, he was the recipient for the "ST Exceptional Patent Award."

Contributors

Paolo Amato is a distinguished member of the technical staff at Micron Technologies. He joined Micron in 2010, where he investigates storage and memory architectures (based on mainstream and emerging technologies) for next generations of mobile systems. He is an expert on statistical methods, error correcting codes, and security, and of their application to mobile systems. From 1998 to March 2008, he was with STMicroelectronics, Agrate, Italy. From 2000 to 2005, he was the leader of the "Methodology for Complexity" team (a global R&D team of about 30 people located in Milano, Napoli, Catania, Moscow and Singapore), a team aimed at developing methods, algorithms, and software tools for complex system management. From 2005 to 2010, he was the manager of Statistical Methods for Non-Volatile Memory-Technology Development group in ST and Numonyx, and he started the development of ECC solutions for PCM devices. Dr. Amato received the laurea degree (cum laude) in computer science from the University of Milano (Italy) in 1997, and Ph.D. in computer science from the University of Milano-Bicocca in 2013. He is the author of more than 40 papers published in peer-reviewed international journals and international conferences, and filed more than 20 patents.

Roberto Bez is presently the senior vice-president of R&D in LFoundry, joined in February 2015. He started his career in the semiconductor industry in 1987 in the Central R&D of STMicroelectronics. The main topics he worked on were the device physics, characterization and modelling, and the non-volatile memory (NVM) technologies, where he contributed to the development of different memory architectures, i.e., NOR, NAND, and phase-change memory (PCM). In March 2008, he joined Numonyx as fellow, driving the development of the PCM and other alternative NVM technology. From April 2010 to January 2015, he has been with Micron as fellow and as process integration director in the Process R&D. He received the Laurea degree in physics from the University of Milan, Italy, in 1985. He has authored or co-authored more than 140 publications and presentations in scientific and technical journals and international conferences and of 60 patents on different microelectronics topics. He has been member of the Technical Committee at IEDM, ESSDERC, EPCOS, and Symp. of VLSI Tech.; he has been member of the ESSDERC/ESSCIRC Steering Committee. He has been contract professor in electron device physics at the University of Milan and lecturer in non-volatile memory devices at the University of Padua, Politecnico of Milan, and University of Udine.

Alessandro Cabrini was born in Pavia, Italy, in 1974. He received the "Laurea" degree (Summa cum Laude) and the Ph.D. in electronic engineering from the University of Pavia in 1999 and 2003, respectively. During his Ph.D., he was involved in the development of solutions for non-volatile semiconductor memories with high-density storage capability. From 2004 to 2011, he was research fellow at the Department of Electronics. Currently, he is assistant professor at the Department of Electrical, Computer and Biomedical Engineering of the University of Pavia. His research interests mainly focus on non-volatile memories and, in particular, on the design, characterization, and modeling of innovative storage devices such as multilevel phase-change, STT-RAM, and flash memories.

His research activities also involve the design of analog circuit for the DC–DC power conversion (especially fully integrated architectures based on charge pump structures), nonlinear electronic systems, and the development of circuits and algorithms for the analysis of electroencephalographic records (EEG) of epileptic patients. He authored about 90 papers in international peer-reviewed journals and conferences and is reviewer of IEEE journal and conferences.

Andrea Fantini received the M.Sc. degree in Electronic Engineering from the University of Pavia, Italy, in 2003. He received the Ph.D. degree in electronic engineering from the University of Pavia, Italy in 2007, with a thesis entitled "Next generation Non Volatile Memories," focusing mainly on the design and architectural aspects of phase-change memories, with Numonyx as industrial partner. After obtaining the Ph.D. title, he held fellow researcher position within University of Pavia and in (2008–2010) with CEA-LETI, Grenoble, France, where he was mainly involved at device level, on the characterization, technological improvement, and reliability assessment of PCRAM and OxRRAM(ReRAM) non-volatile memory technologies. Since March 2011, he is a senior scientist with Memory Device Design Group (MDD) at IMEC, Leuven, where he currently leads the development of OxRRAM technology for embedded applications.

He is author/coauthor of more than 90 papers in international peer-reviewed journals and conferences (including 9 IEDM and 11 VLSI paper). He serves as reviewer of IEEE EDL, TED.

Agostino Pirovano was born in Italy in 1973. He received the Laurea degree in electrical engineering from the Politecnico di Milano, Italy, in 1997, and the Ph.D. degree at the Department of Electrical Engineering, Politecnico di Milano, Italy, in 2000. He joined the Department of Electrical Engineering in 2000, working on the modeling and characterization of transport properties, quantum effects, and physics of Si/SiO_2 interface in MOSFET devices. In 2001 and 2002, he was a consultant for STMicroelectronics working on the development of chalcogenide-based phase-change memories. From 2002, he teaches "Optoelectronics" at the Politecnico di Milano, where he was a lecturer from 1999. In 2003, he joined the Non-Volatile Memory Technology Development Group of the Advanced R&D of STMicroelectronics and he worked on the PCM development at 180-, 90-, and 45-nm node. He is now senior member of the Technical Staff in Micron, working on the electrical characterization and modeling of innovative phase-change non-volatile memories.

Marco Sforzin received the Laurea degree cum laude in electronic engineering from the Politecnico di Milano (Italy) in 1997. In the years 1997 and 1998, he made many tutorials in the area of dynamical system theory at Politecnico di Milano. In 1999, he fulfilled his military obligations and joined the Flash Memory Design Team in STMicroelectronics (Agrate Brianza, Italy) developing SLC NOR Flash memory devices. In 2001, he was appointed project leader of MLC NOR Flash memory prototype for wireless applications and later design manager with the task of bringing MLC devices into production. In 2007, he joined the Advanced Architectures Design Team in Numonyx, developing phase-change memory devices, and in 2013, the Mobile Business Unit R&D of Micron, where he is currently senior member of Technical Staff in Micron. His expertise domains include analog design, high-speed design, full-chip mixed signals design and validation, ECC for memory and storage applications, emerging memory technologies, analytical and statistical modeling. His interests also include information theory, system theory, and neural networks. Mr. Sforzin is the author of 4 international conference papers and more than 20 patents.

Guido Torelli was born in Rome, Italy, in 1949. He received the Laurea degree (with honors) in electronic engineering in 1973 from the University of Pavia, Pavia, Italy.

After graduating, he worked for one year in the Institute of Electronics of the University of Pavia on a scholarship. In 1974, he joined SGS-ATES (which, later, became a part of STMicroelectronics), Agrate Brianza (Milan), Italy, where he served as a design engineer for MOS integrated circuits (ICs), involved in the development of both digital and mixed analog–digital circuits including digital ICs with embedded EEPROM (working silicon: 1978), and where he became Head of the MOS ICs Design Group for Consumer Applications and was appointed Dirigente. In 1987, he joined the Department of Electronics (which then became a part of the Department of Electrical, Computer and Biomedical Engineering) of the University of Pavia, where he is now a full professor. His research interests are in the area of

MOS IC conception, design, analysis, modeling, and characterization. Currently, his work focuses mainly on the fields of non-volatile memories (including phase-change memories) and CMOS analog circuits.

G. Torelli has authored/co-authored more than 320 papers in journals and conference proceedings, one book (in Italian), and eight chapters/papers and is a named inventor in more than 60 patents issued worldwide (59 in the USA). He was a co-recipient of the Institution of Electrical Engineers (IEE) Ambrose Fleming Premium (session 1994–1995). He is a senior member of the Institute of Electrical and Electronics Engineers (IEEE).

Introduction

Roberto Bez

> *Memoria est thesaurus omnium rerum et custos*
> *Cicerone*

Memory has been recognized, since perhaps the beginning, as a key component of human development and civilization. Simply put, without memory we don't exist. For millennia, humans have tried to permanently store information by any means– from carvings in stone to ink markings or to the chemical processing of a variety of thin media–and with various results, depending on time, power and cost. Graffiti as old as 4000 years B.C. can still be seen in some caverns, as well as venerable books, whose endurance is shorter and depends on the preservation method, but extends over many centuries, in many historic libraries. With the advent of electronics, the problem of storing information has been moved from the analog to the digital world, and the adopted solutions are definitely very effective to manage the colossal amount of generated data. So now we can state that semiconductor memory devices have grown, are growing, and will continue to grow in importance because they have become fundamental in practically every electronic system [1].

Indeed, memory devices are key components of all electronic systems, but their function inside the systems has been evolving over time, giving rise to a diversified landscape of devices and technologies [2]. The computer has been the first electronic system to drive the development of memory. The concept of the Turing machine had already introduced two types of memories: a table of rules (instruction memory) and an endless table (alterable data memory). The von Neumann architecture formally unified the two memories, giving flexibility to the program, but introduced the distinction between primary memory and secondary memory (mass storage). That distinction has survived until now, with primary memory being SRAM and afterwards DRAM (with SRAM mainly embedded as cache) and mass-storage moving from hard disk to floating-gate memory, also known as flash. At the same time, the availability of low-cost information storage and processing

R. Bez (✉)
LFoundry, Avezzano, Italy
e-mail: roberto.bez@lfoundry.com

© Springer International Publishing AG 2017
R. Gastaldi and G. Campardo (eds.), *In Search of the Next Memory*,
DOI 10.1007/978-3-319-47724-4_1

has started the trend towards digital storage of all kinds of information, professional as well as for entertainment. A third dimension of memory has been added to data and program memory: mass storage of personal information.

Moreover the explosive growth in the market for mobile phones and other portable electronic devices has been propelled by ongoing decreases in the cost and power consumption of integrated circuits. This is particularly true for non-volatile memories (NVM) based on floating-gate technology [3].

In fact, the dramatic growth of the NVM market started in 1995, and it has been fuelled by two major events: the introduction of flash memory [4–6] and the development of battery-operated electronic appliances, mainly mobile phones. Almost all electronic systems require the storage of some information in a permanent way. The most typical application for NVM has traditionally been program codes for microcontrollers, parameters for DSP's, boot for systems with other type of mass-storage media, data and parameters for security codes, trimming of analog functions, end-user programmable data, system self-diagnostics, etc. [7]. Traditionally, NOR flash technology was the most suitable for code-storage applications, thanks to the short access, read time requested for the execution-in-place of the stored code. Starting from the first years of the new millennium, a novel trend in the NVM market appeared in conjunction with the wide spread of new data-centered applications such as PDA's, MP3 players and digital still cameras. In these portable devices, a huge role is played by the available capacities of the NVM devices to store the data at the lowest possible cost. In this scenario, NAND flash technology entered the market and rapidly became the mainstream technology, even surpassing DRAM in the role of technology driver.

Although the very low cost was the primary driver for the exponential growth of NAND flash technology, in recent years such devices have been able to demonstrate an edge over other technologies also in terms of their performance handling large amounts of data. These capabilities combined with the continuously decreasing costs per gigabyte (GB) to recently create a new market opportunity for flash technology in data-storage applications such as magnetic hard-disk-drive (HDD) replacement. NAND-flash-based HDDs, also called solid-state disks (SSD), provide much better performance than HDDs in terms of sustained throughput, access time, instant-on capabilities and ruggedness.

Nonetheless, efforts continue to improve memory-system efficiency in computing platforms. The usual hierarchy of memories that are used to manage the data manipulation and storage can be optimized with the introduction of new "memory layers" that are intended to mitigate weaknesses of traditional memory systems (costs and power consumption for DRAM, slowness for SSD/HDD).

A very interesting opportunity has been in fact identified for a memory system that could sit in-between DRAM and SSD/HDD, both in terms of latency and costs. Such a memory system, conventionally defined storage-class memory (SCM), represents the ideal realm for NVM technologies due to their low-power capabilities [8]. These kinds of memory have stringent requirements in terms of speed and costs and have driven the search for a new memory concept.

Indeed, at the beginning of last decade, in early 2000, few disruptive technologies had been proposed to replace the standard NVM technology in the semiconductor industry and to enlarge memory applications. In fact, flash had been able to guarantee density, volume and cost sustainability, thanks to a very efficient manufacturing infrastructure and to the continuous dimensional scaling. However, flash miniaturization started to become increasingly difficult with the scaling roadmap moving from planar to 3-dimensional (3D) architectures and with the introduction of new technology-integration challenges. Therefore, there have been opportunities for alternative memory technologies to enter the market and to potentially replace/displace standard NAND flash. Innovative memory has brought the promise of, e.g., faster data storage, lower power, extended endurance, better latency, and so on.

Consequently, strong efforts by companies, universities and research centers have been made to find the best technology able to replace/displace the industry standard. A plethora of new concepts has been presented during the last two decades, spanning more than 30 NVM technologies and technology variations. Among them, only a few are still considered candidates for the displacement role, while most have been already discarded [2, 9].

And, as a matter of fact, the introduction of a new memory concept has not yet materialized to date, at least in terms of meaningful market penetration as compared with NAND flash. Many considerations can be cited to explain why a significant change in memory technology did not happen. But two can be considered the most important. On one hand, all the limitations of NAND-technology scaling have been consistently overcome. The flash industry has had such momentum that it was hard to stop or to change it. There has been a win-win situation between a continuously growing market, requiring ever greater increases in density and reductions in cost, and, on the supply side, the capacity of the industry to generate hundreds of thousands of 300-mm NAND wafers per month. Recently, when planar flash technology scaling stopped at about the 20-nm technology node, vertical integration was already ready to kick in and to provide the expected level of cost reduction. On the other hand, many weaknesses of flash performance, in terms of latency, reliability and so on, have been solved at the system level, while introducing more sophisticated algorithmic and sensing schemes and intensively implementing error-correction management. With this consolidated flash situation and with no a clearly disruptive application able to drive the adoption of a new technology, it has been impossible for any other technology to enter into the big memory arena and become a mainstream parallel to the existing industry standard.

Besides that, any new technology takes a long time to be accepted, since there are many steps to be walked up. First, the concept must be proven; then it must be translated into a memory cell and array, which, in its turn must be integrated within a standard manufacturing line. Moreover, the memory cell placed in a complex array must work by means of a sophisticated algorithm, also needing to be conceived and tested. After the whole technology is validated and as it enters the production phase, it must demonstrate robustness for manufacturing and high reliability to meet the product specifications.

Finally, all of these steps must be concrete enough to be scalable. Only with scalability will there be a reasonable return on the investment needed to set-up the new technology.

This book has the ambitious goal to cover, in some detail, the key aspects in the development of an innovative memory technology, considering various, different emerging memory technologies, but having always flash as a referential example.

Chapter 2, "The Historical Overview of Solid-State Non-Volatile Memory", is a pleasant journey into the development of NVM memory, which happened during the last 30 years. It presents an overview of the key elements in the design and design-technology interaction that enabled the development of EPROM, NOR and NAND flash cell.

Chapter 3, "Emerging Memory Technology", is a comprehensive review of the memory technology and cell concept. It presents the various concepts of memory cell, starting from the flash-cell evolution and the challenges of planar scaling, moving on to the new 3D architecture and describing the most solid candidates to capture some share of the future memory market, i.e., phase-change memory, magnetic RAM, ferroelectric RAM and resistive RAM.

Chapter 4, "Performance Demand for Future NVM", delves into the core of the new memory challenges, i.e., the desired level of performance to fulfill the requirements to move into the memory hierarchy as a storage-class memory. It compares the performance for the various technologies in terms of power, latency, endurance and retention and, based on these factors, elaborates on possible applications.

Chapter 5, "Array Organization in Emerging Memories", addresses one aspect that is as important as the memory concept itself: how to pack together the cells to achieve the highest density required. Arrays are generally made of rows and columns, and a single cell in the array is specifically defined as the crossing point of a particular row with a particular column, which can be intentionally selected. A new memory concept asks for different array architecture and selection modes and so causes different disturbance mechanisms inherent in the functionality.

Chapter 6, "Data Sensing in Emerging NVM", touches one of the most important elements in concept development: how to sense the physical state of the cell and translate it into a digital signal. Here, the sensing circuits used in various emerging memories are described and compared with the sensing principles of flash and DRAM to underline the differences between them and the emerging techniques.

Chapter 7, "Algorithms to Survive: Programming Operation in Non-Volatile Memories", has embedded its title the crucial concept: algorithms are needed to keep a technology alive. Together with data sensing, without the proper algorithm that monitors the cell functionality, but also the array organization, no memory concept will survive. It is very important then to review the algorithm used for flash and then to compare the ones proposed for the emerging technologies.

Chapter 8, "Error Management", explains the various techniques adopted for error-correction management in the memory. As already mentioned, this has been a great breakthrough that has prolonged the life of the industry standard memory, especially NAND flash, and it is considered fundamental for any new memory technology that would like to play a primary role. An interesting historical overview of various possible error-correcting code precedes the analytical description of innovative techniques.

Chapter 9, "Emerging Memories in Radiation-Hard Design", briefly deals with the specification of radiation hard memory and compares the intrinsic advantages of the emerging memory with the flash one.

It is hard to predict the future shape of the memory landscape. But it will be harder to imagine it without comprehending the basic know-how that has been generated so far and the having an accurate perspective on memory development, from flash to the recently proposed emerging technologies. This book will help readers to better understand the present and future memory scenarios.

References

1. S. De Boer et al., "A Semiconductor Memory Development and Manufacturing Perspective", ESSDERC, Venice, 2014.
2. L. Baldi et al., "Emerging Memories", Solid-State Electronics, v. 102, December 2014, pp. 2–11.
3. R. Bez and A. Pirovano, "Overview of non-volatile memory technology: markets, technologies and trends", in "Advances in Non-volatile Memory and Storage Technology", edited by Y. Nishi, Woodhead Publishing, 2014.
4. "Flash Memories", edited by P. Cappelletti et al., Kluwer Academic Publishers, 1999.
5. R. Bez et al., "Introduction to Flash Memory", Proceedings of the IEEE, vol. 91, n. 4, 2003.
6. "Nonvolatile Memory Technologies with Emphasis on Flash", edited by J. E. Brewer and M. Gill, IEEE Press, Wiley & Sons, 2008.
7. G. Forni et al., "Flash Memory Application", in "Nonvolatile Memory Technologies with Emphasis on Flash", edited by J. E. Brewer and M. Gill, IEEE Press, Wiley & Sons, 2008.
8. R. F. Freitas and W. W. Wilcke, "Storage-Class Memory: The next Storage System Technology", IBM Journal of Research and Development, vol. 52(4/5), pp. 439–448, 2008.
9. K. Prall et al., "An Update on Emerging Memory: Progress to 2X nm", International Memory Workshop (IMW), Milan, 2012.

Historical Overview of Solid-State Non-Volatile Memories

Giovanni Campardo

There are 10 different kinds of people.
Those who understand binary, and those who don't.
(From the web).

1 The Story

I can say that my career can be superimposed, for a long time, on the story of non-volatile memories. I am not so old that I witnessed the birth of non-volatile memories, but, when I started to work on integrated circuits, the memory-cell dimension was in the micron range: 1.5 micron (μm) as the minimum length and the size of the device was 512 Kb.[1] The contact size was 1.5 μm by 1.5 μum. Today the minimum size is in the range of a nanometer (nm) and the size of the device is approaching a terabit (Tb).

At that time, there were some fixed points that defined a 'good' device:

- Chip size must be around 50 mm.[2]
- Access time, in read mode, must be around 100 ns.
- All the bits must be 'good', i.e., no failure bits are permissible in the parts.
- Current consumption must not be greater than 50 mA.

We will see how these 'fixed points' changed during the years, driven by the applications.

The first commercial non-volatile memory with a solid-state technology, was an EEPROM, the Electrical, Erasable, Programmable Read Only Memory, at the end of the '70s, but, at that time, I already was in college. I was first involved in the EPROM development, Fig. 1.

---

[1]Memory-hardware designers always mean bit, speaking about memory size. So, for me, a memory of 1 M means one megabit. Software people always speak about Byte!.

G. Campardo (✉)
ADG Automotive Digital Division - Product Design, STM, Agrate, Italy
e-mail: giovanni.campardo@st.com

© Springer International Publishing AG 2017   7
R. Gastaldi and G. Campardo (eds.), *In Search of the Next Memory*,
DOI 10.1007/978-3-319-47724-4_2

Fig. 1 EPROM device with
the silicon window that
enabled erasure by UV

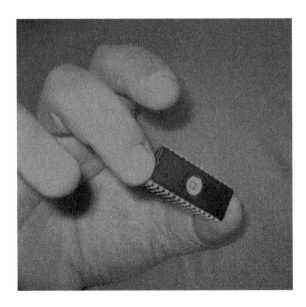

Fig. 1 EPROM device with the silicon window that enabled erasure by UV

The need to have a device able to maintain data stored inside it, also when the voltage is removed, was a fundamental requirement, as well as simple ways to store and retrieve the data and programs.

In the 1980s, there were no MP3 music players, no digital photo cameras, and neither any digital televisions, cellular phones, tablets, portable computers... so, why did we need nonvolatile memories in the 80s?

All electronics machines, like desktop computers or industrial machines like the automatic coffee maker, the dishwasher, the ice cream machine, etc., need a program to work. Perhaps hard to believe, but, in the 1980s, the washing machine, for example, had a camshaft, like a carillon. A metallic cylinder with variously geared teeth, the cylinder rotated slowly, and the teeth opened and closed various electro-mechanical relays.

Of course, the need to reduce the size, decrease the power consumption, and increase the 'intelligence' of the executive program imply having the electronics on board and, most importantly, having storage to encode the program to be executed.

In an EPROM, you can write the bytes individually but the erasing is done by removing the chip from the board and putting it under a UV light for at least 20 min. After that, you must put the EPROM into the socket of a 'dedicated machine' that is able to write the memory code and verify it. Finally, one reinserts the EPROM into the devices' application board with a new program inside.

Of course this is not acceptable today! Can you imagine doing the same now for all your your devices, e.g., for your cellular phone? Each time you want store a new phone number, you have to turn your phone off, open the case, remove the memory, employ a 'dedicated machine' where your address book is updated, and then put the memory back into the case, etc.

If you want to save an SMS, you have to do the same operation each time that you decide to write a new message. It would not be practical.

EPROM has a glass window in the package to permit erasing the device. The package is made of ceramic and not plastic to increase robustness, but the drawback is the added cost, due to the presence of the window. The mechanism to program is called a Channel Hot Electron (CHE), and the erasing was done by tunneling, produced by UV radiation.

The circuitry for an EPROM device (Fig. 2) can be rendered into these fundamental blocks: the decoders for rows and columns; the sense amplifier; the circuitry to read the status of the cell; the circuitry for the redundancy; and the registers reset by the EPROM cell called UPROM, used to store the fail addresses; at every read request, the circuitry compares the address presented to the redundancy registers. If

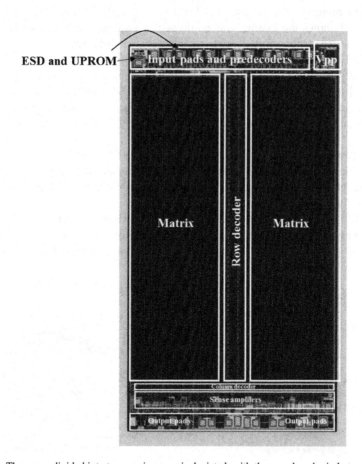

Fig. 2 The array, divided into two semi-arrays, is depicted, with the row decoder in between. The column decoder is located above the sense amplifiers to read and the output pads. In the top part are the input pads and predecoders for both rows and columns. Finally, the VPP pad, ESD, and UPROM are positioned as shown

the address is recognized as one of the failed, the read is redirected to the redundant cells, otherwise 'normal' cells will be activated. In the device depicted in Fig. 2, the UPROM (Unerasable Programmable Read Only Memory) was inserted into the input buffers, so the comparison was done very early in the access time of the redundant parts, which was not so different from the good one. The UPROM structure was really very clever; he who invented this, must have had a moment's glance at 'the God notebook'. The UPROM was performed by the EPROM cell but, the problem is that these cells must not be erased when the user decides to erase the device by UV light. So, this cell was covered by metal and the last layer was to connect the transistor but to avoid the possibility that the light could reach the cells, like a waveguide. To this end, the gate connection was designed not like a straight line but like a wavy line.

Other circuits present were the ATD generator, the Power On, the I/O buffer with the ESD circuitry, and control signals PAD, such as the CE# and the OE#.

Figure 3 shows the biasing voltage needed to program (or write) an EPROM cell, on the left. On the right is a photo of the real physical structure.

The ATD, Address Transition Detector (Fig. 4), was what we can call a 'pseudo clock'; EPROM had no clock but, as we can understand, a synchronization circuits is needed to scan the sequential operations, so the ATD observes the change of address and also the control signal. When a change occurs, a pulse was generated, and this pulse was used as a 'start' to drive the event sequence of the read chain blocks.

The Power On circuit, many times called POR (Power On Reset), is a very simple circuit able 'to detect' the supply voltage ramp, and, during it, to generate a pulse to set the necessary circuits into a known state, or, more simply stated, a hardware reset.

The ESD circuitry is used to prevent possible overvoltage or under, beyond the limits to avoid the possibility of destroying the circuitry. ESD circuit will turn on before the discharge will be able to reach the first transistor gate. These structures are inside the PADs.

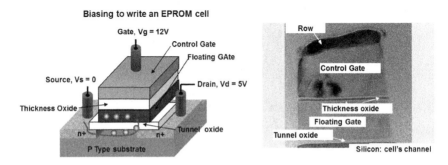

Fig. 3 EPROM-cell write biasing

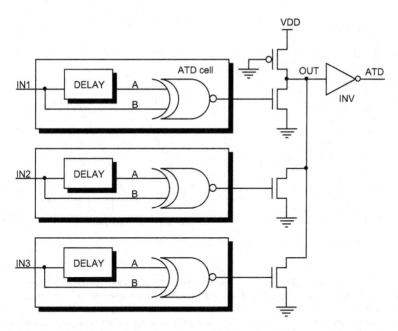

Fig. 4 The block in the box is able to generate a pulse when the IN# changes. The composition of several control blocks enables detection of the transition of the input signals. The various outputs are combined by means of a distributed NOR

The control signal is indicated by CE#, the Chip Enable#, and active low (# is the symbol to indicate that the signal is active when it is low, normal ground, 0 v). This convention, to have the active state at low-voltage value of the input, was due to the NMOS circuitry. Low input means: no consumption, the NMOS transistor is off, and a sure, full value exists of the output of the first inverter. This signal turns on the device while the OE#, Output Enable# removes the output buffer from the three state situation (or high impedance state) where the Byte value was already loaded.

So, this was really 'poor' circuitry as compared with the circuitry we have today in a non-volatile memory. In fact, in the '80s, the process was NMOS with no PCH transistor available! Simple process, only NMOS transistors, means complex circuit, if you want to have good performances.

Bootstrap techniques[2] are adopted throughout, otherwise the voltage node could easily reach the ground value. To reach the VDD, NMOS limit, i.e., the capability to reach the VDD due to the positive threshold, you need to set the gate of the generic NMOS a step above the VDD.

---

[2]For me the word bootstrap didn't have a clear meaning. I knew what the bootstrap is, but when I eventually understood the real meaning of this, and I started to imagine someone trying to lift up himself pulling on the boot straps, I really came to understand the meaning of this in electronics circuits.

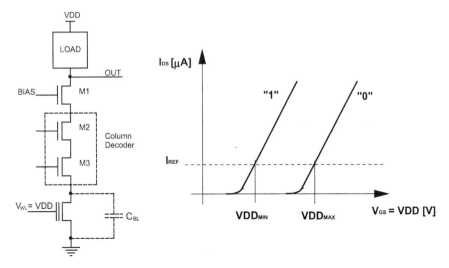

Fig. 5 The first sense amplifier made a simple comparison with the fixed current represented by the LOAD. Logic "1" and "0" differs from a threshold shift, obtained by a program. The VDD space was the voltage on the gate row. The space in between VDD_{MIN} and VDD_{MAX} defined the VDD voltage range to be read. In that space, a cell with a current greater of I_{REF} is recognized as a "1", and with a lower current as a "0"

The sense amplifier to read was, for the initial devices, to a simple 'inverter stage' (Fig. 5), written cell sink '0' current, virgin cell sink current.

The read mode circuit was like an inverter.

In the latter half of the '80s, devices started to have the first differential sense amplifiers that used a reference cell (Fig. 6).

After a while, a great revolution that has lasted 30 years began. The era of the mobile would eventually arrive, and nothing could stop it.

I remember, like it was yesterday, when the boss of the R&D (he had to be one of the contributors to this book) entered our open workspace, where the memory group was sitting, with a sample in his hands. He was very, very happy, and he said 'they erase.....'; he was actually saying that they are trying to erase an EPROM device, a test device, using an electric pulse, not like an EEPROM but to erasing the entire matrix. One of my colleagues, a test engineers, answered: "…we call these rejects…!".

I will never be able to forget the face and the look of the R&D boss, a very perturbed manager!

In any case, a story that I told to the student during some seminars at the university recounted how Flash memory was invented to make it possible to erase, in one shot, all memory content. This request came from a military application. They wanted to have the possibility to destroy the software, or firmware, should a missile or airplane or whatever contains a piece of proprietary software that could fall into the hands of the enemy.

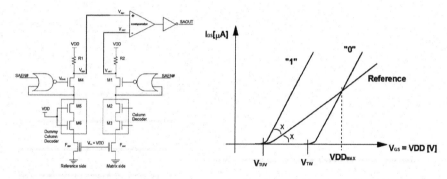

Fig. 6 A differential sense amplifier means having two branches to compare one to the other. This speeds up the read phase. In the first sense amplifier approach, the main issue was to enlarge the VDD working window. So, using a cell to trace the process and temperature variability, a sense must be able to have a reference characteristic in between the "1" and the "0". This was obtained with unbalanced loads R1 and R2

Fig. 7 Flash cell erase biasing

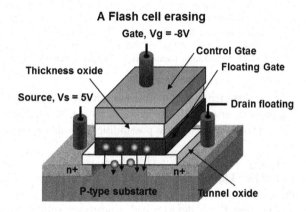

Whether or not the story is true, the possibility to modify the content of the memory without removing it from the application board was the first requirement. And so, why not use EEPROM? EEPROM was and is a non-volatile memory, programmable and erasable by electrical application, Fig. 7. With this, there is no need to remove it from the board.

The price to be paid was the size of the cell. Each cell, or each byte, could be addressed independently and, using high voltage, an electric-tunnel erase is achieved. While this was a great advance in term of flexibility but without the ability to have mass storage memory, only a few kilobytes are possible.

Now having the capability to modify the memory content using external commands, the market that seems to govern our lives (is it true?) Started to ask for more and more.

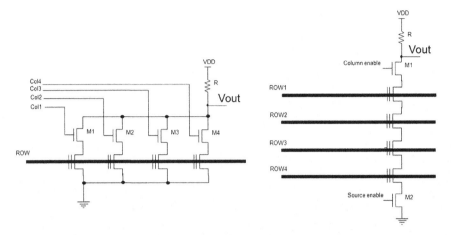

Fig. 8 Let me skip forward for a moment. On the *left side*, we have the NOR organization. Each cell connected to the same ROW will be activated. If the correspondent column decoder transistor will be ON, the node Vout will be driven to the relative voltage. NOR organization means: it is enough to have only one element turned on to produce the output. On the *right side*, the NAND organization: to drive the output, all of the cell must be activated. The different values of the voltage ROW# will define the cell that must be read and the cells that must be turned on to be a pass. NAND organization means: all the elements must be turned on

As a first step, when the memory devices were still EPROM and not yet Flash, the process, also for the memory, became CMOS. The memory designers, at least, could exploit a lot of new possibilities but also there were a lot of new things to learn.

Firstly Flash, NOR architecture, is not an acronym, like for EPROM or EEPROM; Flash means 'in a flash', like a lighting. NOR is based on the matrix organization, see Fig. 8.

The first Flash devices had a structure similar to the EPROM: very simple circuitry, no internal algorithms machine, programmed with CHE, and erasure by tunneling.

Now came the era of the ASM, Application Specific Memory or memory design for a special application. For example, a digital filter, an FIR, constructed with an EPROM, used like a look-up table to speed up the calculus and was used in the first digital television, Fig. 9a; and an ASIC, with a macrocell Flash, was used for automotive application, Fig. 9b.

Another ASM was an I/O port expander for automotive application, composed of a Flash and a SRAM memory on the same die, see Fig. 10. Each time the car was turned off, the parameters written in the SRAM were stored in the Flash, ready for the next ignition. At that time, for many different types of applications, a software program was introduced to recover the depleted cells.

We can say that the Flash story is the story of the erase mode and the way to recover the problematic information caused by the erase. For an EPROM, the distribution of the virgin cell is tight. The light erasing is able to set the floating gate

(a)   **(b)**

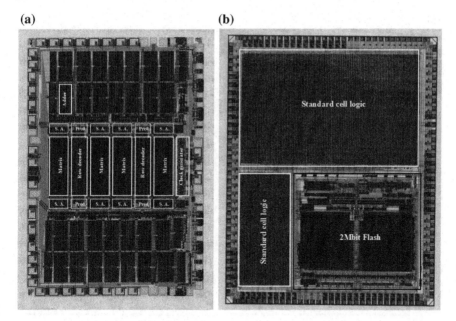

Fig. 9 **a** A FIR filter using an EPROM memory, divided into four semi-arrays to speed up read. The content of the memory is a look-up table whole outputs are added by a fast adder to produce the digital processing that realized the required filter. In this device, the column decoder was not present to speed up read. **b** ASM memory chip. 2 Mbit Flash memory inside a pad limited device with logic, designed with the standard cell technique. Note the regularity of the logic cells compared with the part of memory realized by designing custom transistors

to the original situation of charge neutrality, so that the distribution of the erased cells is only dependent on the structural differences between the cells. For a Flash, the erasing is a collective action done by electrical pulses. The results is a large distribution of the threshold voltage, Fig. 11.

Another of the points that distinguish EPROM from Flash is the reference cell and also the matrix organization that is specifically related.

In an EPROM device, the matrix is in a continuum with the reference cell, inserted in the matrix as a reference column for each output.

This solution is feasible because the erasing is done by light and results in a tight distribution. Since there is a cell for each output, if the matrix is organized with eight outputs, we will have eight reference cells for each row. This structure will be repeated for each column. This structure has the highest similarity to the cells matrix in terms of process and temperature variation, leaving the best solution for the differential read mode. This solution, for a Flash, is not more likely, because, if the reference cells were inserted inside the matrix when the erase is performed, the reference cells will also be erased and the distribution is also be larger in respect to the UV distribution. It will be possible to reserve 'special' blocks for the reference cells, but a lot of space will be consumed. The final solution was to remove the

(c)

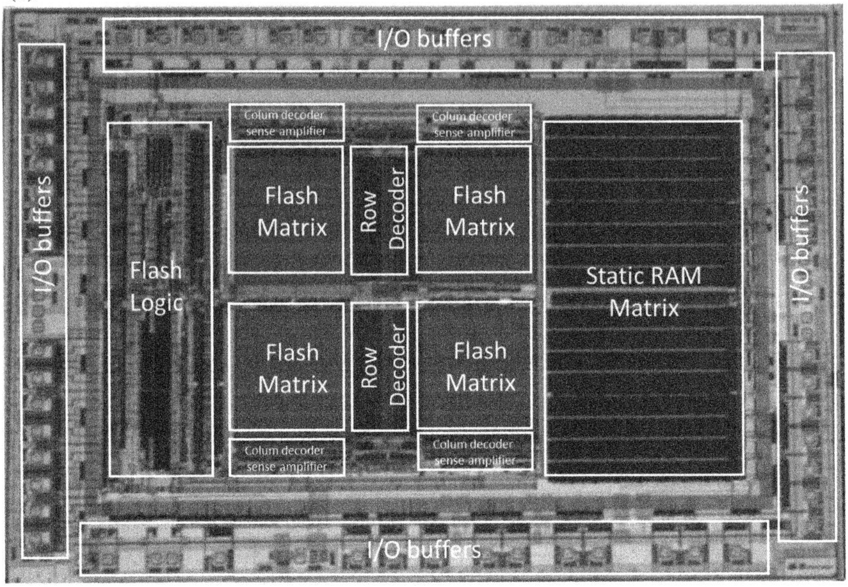

Fig. 10 An I/O Port Expander done with a 1Megabit Flash, plus a 16Kbit SRAM, to store the program and data for the automotive application

reference cells from the matrix and create another little matrix, outside the matrix space, while inside all the reference cells remain that one might need.

The second point was the fact that the EPROM was a 'single' matrix, a single monolithic block because it is not possible to 'hide' from the light a portion of the matrix and to unveil it when we want.

Anyway, the first Flash device did not have a matrix divided into sectors. The matrix was a single block, too, like an EPROM, but with the possibility of erasure by electrical signal. This was a great achievement, having the possibility to erase by electrical signals without also remove memory from the application board.

For all these reasons, the design of a Flash-memory array is more difficult compared to the design of an EPROM-memory array.

For the EPROM, the matrix structure is designed focusing on the access time, which placed constraints on the row and column length. This justifies splitting the matrix into different arrays with shorter rows and columns, by multiplying row and column decoders, called sectors.

At that time, there occurred no great circuitry revolution. The erasing was achieved by driving all the rows to 0 V, all the bit lines floating, and the source connected to a high voltage, typically 12 V.

The only further change to be accomplished was due to the fact that, in an EPROM, the row decoder that addresses logic always guaranteed a row at read voltage; this was one of the problems for the reliability of the gate stress generated;

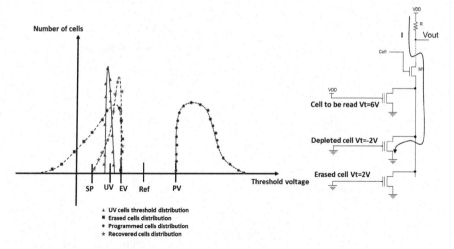

Fig. 11 The electrical erase produces a large distribution. Starting from a written cells distribution, all the cells have a threshold voltage greater than a defined value, called Program Verify (PV). The erase is done by an electrical pulse to remove electrons from the floating gate, so that the erased cells will have a threshold less than the Erase-Verify voltage (EV). The final distribution could also have cells with a threshold voltage less than than 0. These cells, called depleted cells, can produce a false read. The scheme on the right represents a bit line. We would like to read the first cell, a programmed cell with a 6 V of threshold. No current must flow from the load to the ground, but the cell with a threshold of −2 V sinks current also if it is not addressed. The results, for the sense amplifier, will be the same as if a logic '1' is present. Soft Program algorithm is able to detect a cell with a threshold less than SP voltage, and, with well-defined pulses, to recover these cells with a well-controlled program. It is fundamental not to move the other cells, already erased. The final erased distribution will be between the SP and EV voltages. The Ref value is the reference voltage threshold of the reference cell used to read. The UV distribution is graphed, which has very tight distribution around the UV threshold voltage

in any event, to erase we need to have the gate at 0 V. So a new signal was introduced to force all the gates to ground, Fig. 12.

Figure 13 shows two possible sector architectures. The first, on the left, shows different i.e., a portion of the matrix with different sources nodes to be erased separately, with a common row. On the right, are again different sectors with different sources, but the row is not common to the sectors, and the columns are inserted separately with a local column decoder between them. This solution can avoid any stress on a sector not involved during the modification, program, and erase modes.

The main problem was the spurious consumption during the erase. The erase is performed by a tunneling effect, so the current used is really very low, but the high voltage applied implies a spurious current, the Band-to-Band current, which is an order of magnitude greater than the tunneling current. The market pushes forcefully to design mobile devices that have the characteristics of low-power consumption and long battery duration, so that means a low supply voltage.

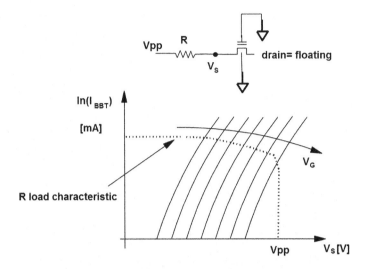

Fig. 12 The first electrical erase was performed by a high voltage, typically 12 V, on the source, which is the common node of the sector. To decrease the spurious effect, the Band-to-Band (BBT) current, a resistor, R, is connected between the external voltage Vpp and the source matrix to exercise a negative feedback and limit the spurious current. The graph represents the BBT current for various gate and source voltages V_G, V_S. As we can see, the R characteristic ranges from a high current to 0 as the erase proceeds

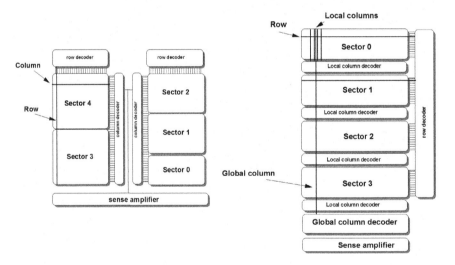

Fig. 13 Two examples of sectorization matrix array

It was not possible any more to have a dual-voltage device with a core voltage equal to 5 V and a modify voltage equal to 12 V, exactly like the EPROM.

The solution was to decrease the supply voltage and remove the 12 V from the board. Thus, the single supply-voltage devices was created, 5 V for all.

But, how it is possible to erase having only an external 5 V supply?

This approach is feasible because it is true that the power consumption, per cell, was around, during Program, 1 mA but we need high voltage on the row, typically 10V with no power consumption because it is an RC charge while, for the drain, 5V is good enough to program. The same for read, but, for the erase pump circuit, diode and capacitance typology could be designed although the output resistance of these circuits cannot reach the value we need to sustain 12 V at a few mA of current consumption. To do this, we would need larger capacitances that consume too much space.

Technology helped to find the solution. While erasing needs to have an electric field between the control gate and the source, this could be reached not only by putting all the voltage on the source, but also by splitting the voltage between the gate and the source nodes.

The solution was to put a negative voltage on the gate and a positive voltage on the source, typically -10 V on the control gate and 5 V on the source, to achieve the electric field needed for the tunneling. Always driven by the mobile-phone requirement, sectorization explored other solutions. Of course, the sector, the part of the matrix erasable without interference with the other sectors, must have the highest granularity as possible, but this conflicts with the die area, the real parameter to be verified to produce a salable device. Another point was the problem related to the time needed to erase. Typical time-to-program is in the order of microseconds (μs), while to erase the order of magnitude is second (s). Of course, while the memory is replacing its content, with a programming or an erasing or both, the microcontroller in the system will be stopped, waiting for the memory-activity conclusion; this could be not acceptable.

The 'read-while-write' and 'read-while-erase' concepts were born to solve the problem. This was one of the most important concepts for the cellular mobile application during the first years after the introduction of cellular phoned.

Let's come back now to the new erase voltage distribution with the negative voltage applied to the gate.

The principal issue was the incompatibility of the negative voltage with the NMOS transistors.

With the final row driver as a CMOS inverter, the PCH transistor will switch the positive voltage on the row and the NCH transistor on the ground. A 'normal' NMOS cannot switch a negative voltage because it has the substrate biased to ground, and the reverse diode, between the source (and drain) and the substrate, cannot sustain more than a threshold voltage before starting to conduct current.

So, the first solution was not to change the process, but to invent a circuital solution to overcome the problem, Fig. 14, which depicts the solutions.

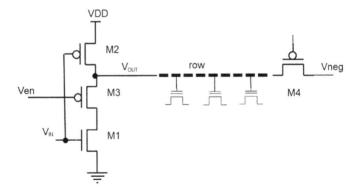

Fig. 14 The circuital solution to solve the problem related to the impossibility of putting a negative voltage on the NMOS channel terminals

The negative voltage will be switched on the matrix row from a PMOS put on the other end of the row, in respect to the row-decoder position. The row contacts never see a NMOS transistor but always PMOS. In this way, no direct diode current will be activated.

This solution was a great one, but not so economical. The row-decoder volume will greatly increase, and the necessity to use only PMOS transistors to manage negative voltage implies also other limitations. In the single voltage device, we need to generate internally the voltages greater or smaller than the supply voltage, using the pump circuits.

To generate negative voltage, as for the row decoder, we cannot manage negative voltage, so the negative voltage pump must be composed of only a PMOS transistor. This choice limits the pump efficiency because PMOS charge, the holes, are heavier and slower with respect to the NMOS charge, the electrons.

The next step was the introduction of the triple well, Fig. 15.

This process step enables having an isolated p-type substrate, with a buried layer and well on the lateral side. The NMOS transistor diffused inside this 'isolated' well could bias its relative ip-substrate and avoid any diode direct current.

Also for the read operation, a low-voltage supply introduced many diverse solutions. For the early EPROM. the voltage put on the matrix row, the cell control gate, was the VDD. With the low-voltage solution down to 5 V, and then to 1.8–1.6 V, it was mandatory to generate internally the voltage to read, using again a pump circuit.

A simple comment readily becomes clear: the Flash story correspond to the erase story, and this it is true, of course, that Flash was born to enable the electrical erasing and thus the requirements of the mobile market.

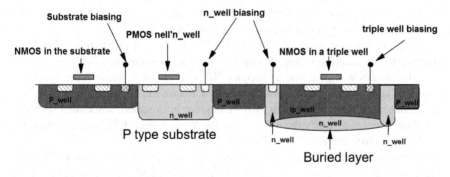

Fig. 15 The triple well technology

The technology has followed the well-known Moore law,[3] and the cell shrink path, too. Over 30 years, the minimum-size dimension has evolved from micron to nanometer.

Over those 30 years, a nonvolatile memory-chip size, in term of cells number, evolved from 64 Kb to 128 Gb, a factor of 2,000,000 greater.

This change reduced the cell size, but also the number of electrons stored in the floating gate. The first EPROM had, in the floating gate, around 50,000 electrons (this is an estimate based on the capacitance calculation and the voltage related), while today we have a few tens of electrons.

This, with the complex technological steps and the large number of cells in each device, has resulted in a very high possibility to have fail cells in the array. Moreover, with the usage, the cycle of program and erase also generate aged cells that must be recovered or substituted for during normal life of device applications.

That was the time of the Error Correction Code (ECC), and let me make a few comments about this concept.

During our normal, everyday life, we use many devices that increase the quality of life in this society, i.e., the information society. Our cellular phones, with a memory capacity and a working frequency comparable to a computer, are used not only to make calls, but also to listen to music and the radio, to watch movies, and to send and receive other messages. Moreover, we play electronic games, we maintains a calendar, an agenda, and an address and phone book, etc.

At every turn, we are exploiting the laws of quantum mechanics. Every time we record an address or a cellular phone number or we save an SMS, we use the tunnel effect to erase and write data (depending on the type of nonvolatile memory we are using), and, if we want to find a street using our GPS, we are using relativistic

---

[3]In 1965, just a couple of years after Jack Kilby patented the bipolar transistor, Intel's co-founder Gordon Moore observed that the number of transistor per square inch in an integrated circuit doubled every year. Moore foresaw that this trend would be valid for the years to come: indeed, the doubling of the density of the active components in an integrated circuit proved to occur every 18 months, and this is the commonly accepted value for the definition of Moore's law today.

mechanics to optimize the information. All these operations require use a large data exchange, and this exchange must work without error.

I tried to explain to my fried what the ECC is: it is exactly when a wife send her husband to the market to shop. While first she explains clearly what you have to buy, she always will give you a shopping list in a written form. In more synthetic form, this is redundant information. Although I always fail to successfully complete the errand, a transmission cannot fail. Redundant information, associated with the initial one, serves to verify the accuracy of the transmission. ECC needs more cells inside the matrix, increasing the die area.

Then came a really fundamental change, the multilevel introduction, Fig. 16. The possibility to be able 'to count' the electrons displaced, during the program, in the floating gate.

Of course, we are not able literally to count, but we are able to discriminate the value of the cell current and, knowing that this current is related to its threshold, to detect 'where' is the threshold of the cell. In Fig. 17, we see on the left the distribution for a single-level device. The thresholds are divided in two distributions, the "1" and the "0" logic. A reference put in the middle is able to understand cell thresholds and to compare the reference current and the matrix cell current.

If we are to be able, through the program action, to program a cell at the threshold level we want, as in the right-hand side of Fig. 17, we can have more than two distributions. For a 2bit/cell, we will have 4 distributions with logic values "11", "10", "01", and "00". Three different references can enable comparisons between cell current and the references that define the distributions to which the cells belongs.

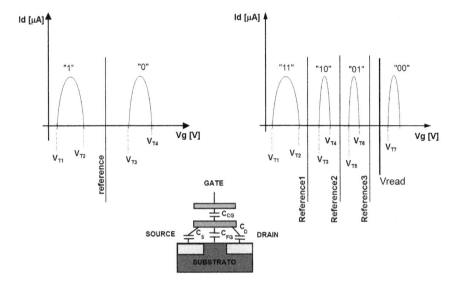

Fig. 16 The threshold distribution (*left-hand side* for a single level and *right-hand* for a 2bit/cell)

The shrinking dimensions of the technology continues, but the market requirements are ever more aggressive. Our times require tons of memory to record all the moments of our lives. The next trick to satisfy the customers was the use of the z-dimension.

A way called 'more than Moore' enables the placement, in the same package, of more dice, one piled on top of another, as shown in Fig. 17. Today, the stack up can reach up to 16 levels.

In Fig. 17, on the left is a wafer that was ground with a mixed process—chemical and mechanical—to obtain a 70-μm thickness starting with 800 μm for the original wafer. With this thickness, the wafer became elastic. The dice grinded are attached in a stack and wired to a substrate that works like a board to connect the wire bonding to the external connections, composed of metallic balls.

All USB pen drives, the Solid-State Disk (SSD), and memory used in the automotive dashboard, or inside a cellular phone, are done with this technology.[4]

While the customers want an increased complexity of the system, the management of the complexity must be handled within the system and not inflicted on the user, of course.

The next generation was the Managed-Memory generation. We already noted that the technology shrink implies an increase of defects, and how the solution was the introduction of ECC. But the user wants system compatibility. When you connect your external device to the PC, it is recognize as a peripheral and, on your screen, you will see a folder, like for all the other files you have on your internal hard disk.

Moreover we also need to build all the algorithms needed to extend the lifetime of the device, like the Bad Block Management, which is a way to discover if a block is beginning to fail and then to substitute some redundant block for it, without

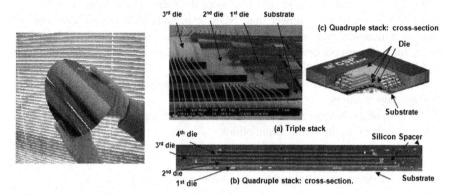

Fig. 17 Stacked technology

---

[4]My first PC, a few years ago, after the Z80 and some other machines, was a 16 MHz clock PC with a 16 MB hard disk. Today, in a cellular phone, the nonvolatile memory is measured in GB.

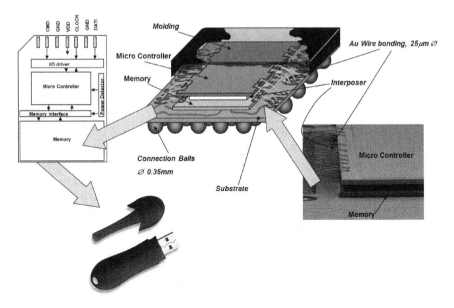

Fig. 18 A managed memory, a CARD

any loss of information. To do this, a microcontroller is embedded inside the stack, to relieve users from the having to control the memory, Fig. 18.

Up until the last ten years, the floating-gate memory was Flash NOR, but memory Flash NAND then started to become the leader. The compactness of the matrix layout, but above all the choice to use the tunneling effect also for the program,[5] provided the possibility to write to many cells as compared to the NOR solution that uses CHE. This means low-current operation, but also the possibility to increase the communication speed between the memory and system.

NAND is intrinsically slower than NOR, especially in read. We can state an order of magnitude, nanosecond for the NOR and microsecond for the NAND but, again, for a power-consumption reason, NOR reads 256–512 cells at the same time and NAND reaches 16,000.

So, NAND enhanced the performance of the system and became the real mass-storage memory.

The size of the NAND die became greater over the years, now more than 1 cm^2. Coming back to the initial assumptions:

- Chip size must be around 50 mm^2

- Memory device is today greater than 1 cm^2

  - Access time, in read mode, must be around 100 ns

---

[5]This was not a 'choice' but, due to the array organization in a NAND structure, it is not possible to apply a current to program with the CHE effect.

- Memory device access time is in the range of microseconds. The system 'solves' the problems

  – All the bits must be 'good', no failure bits are permitted inside the parts

- Memory devices are sold with a maximum value of the failed blocks permitted

  – Current consumption must be not greater than 50 mA.

- System power consumption could be very high, especially for the interface working at high frequency.

And now? People are exploring the z-dimension, memory, NAND memory, with the cells stacked one over the other, to save space and increase memory availability. What are we likely to see in the next future?

References

1. G. Campardo et al. *"A 40 mm² 3 V 50 MHz 64 Mb 4-level Cell NOR Type Flash Memory,* IEEE Journal of Solid State Circuits, Vol. 35, N0 11, November 2000.
2. G. Campardo and R. Micheloni *"Architecture of nonvolatile memory with multi-bit cells"* Elsevier Science, Microelectronic Engineering, Volume 59, Issue 1-4, November 2001, pp. 173–181.
3. G. Campardo, R. Micheloni *"Scanning the Issue",* Proceeding of the IEEE, VOL. 91, No. 4, April 2003, Special Issue on the Flash Memories.
4. G. Campardo et al. *"An Overview of Flash Architectural Developments"* Proceeding of the IEEE, April 2003.
5. R. Micheloni et al. *"The Flash Memory Read Path: building blocks and critical aspects"* Proceeding of the IEEE, April 2003.
6. G. Campardo, R. Micheloni, D. Novosel *"VLSI-Design of Non-Volatile Memories",* Springer Series in ADVANCED MICROELECTRONICS, 2005.
7. P. Pulici, G. Campardo, G. P. Vanalli, P. P. Stoppino, T. Lessio, A. Vigilante, A. Losavio, G. Ripamonti *"1 V NOR Flash memory employing inductor merged within package"* Electronics Letters, vol. 43, no. 10, Page(s): 566–567, 2007.
8. G. Campardo, R. Micheloni *"La rivoluzione delle memorie,"* Le Scienze, italian editon of SCIENTIFIC AMERICAN, number 468- August 2007, pagg. 104–110.
9. R. Micheloni, G. Campardo, P. Olivo, *"Memories in Wireless Systems",* Springer Verlag, 2008.
10. A. Maurelli, D. Beloit, G. Campardo, *"SoC and SiP, the Yin and Yang of the Tao for the new Electronic Era",* Proceeding of the IEEE, Vol. 97, number 1, January 2009, Special Issue on the 3D Integration Technology.
11. G. Campardo, G. Ripamonti, R. Micheloni *"Scanning the Issue",* Proceeding of the IEEE, Vol. 97, number 1, January 2009, Special Issue on the 3D Integration Technology.
12. G. Campardo, F. Tiziani, M. Iaculo *"Memory Mass Storage"*, Springer Verlag, March 2011.

Physics and Technology of Emerging Non-Volatile Memories

Agostino Pirovano

1 Challenges in Floating-Gate Memory Scaling

Starting from their original introduction at the end of the 1980s, the fundamental properties that enabled flash memory to become first the non-volatile mainstream technology and later the semiconductor technology driver for scalability also enabled flash to follow the well-known Moore's law for semiconductors. This capability was largely demonstrated by the NOR flash-technology evolution that has followed that of the standard CMOS, introducing into the basic process flow many of the materials and modules already developed [1, 3]. In particular, considering the requirements for fast random access time, the supply voltage for mobile application down to 1.8 V and the efficient programming and erasing algorithm execution, starting from the 180-nm technology node, the CMOS structure has mainly followed the high-performance logic roadmap [4–6].

Figure 1 depicts the NOR-flash cross sections for successive generations, from the 0.8 um to the 65 nm, with the materials and the basic modules that have been introduced for each generation. With the device scaling, the front-end process module took advantage of improved salicidation processes introduced in high-performance logic, and the gate material has evolved from WSi_2 to $TiSi_2$ and finally to $CoSi_2$. Similarly in the back-end, the metallization evolved from single aluminum to triple copper layers.

Figure 2 shows the cell-size reduction as a function of the technology node F, where F represents the minimum feature size. Solid points represent the real cell sizes for NOR- and NAND-flash technologies put into production in the last 15 years, clearly showing the capability for both architectures to follow the area scaling suggested by Moore's law. In a flash-NOR cell, the theoretical size is 10 F^2,

A. Pirovano (✉)
Technology Development, Micron Semiconductor Italia,
Via Trento, 26, 20871 Vimercate, MB, Italy
e-mail: apirovan@micron.com

© Springer International Publishing AG 2017
R. Gastaldi and G. Campardo (eds.), *In Search of the Next Memory*,
DOI 10.1007/978-3-319-47724-4_3

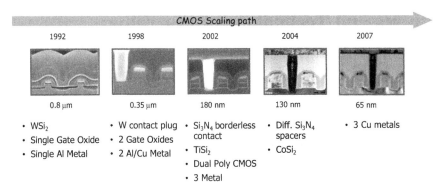

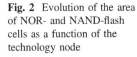

Fig. 1 The NOR flash-cell cross section for successive generations, reported in terms of technology node (year of production) and key materials introduced

Fig. 2 Evolution of the area of NOR- and NAND-flash cells as a function of the technology node

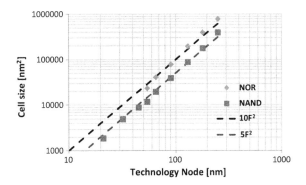

while, in a NAND cell, the size is around 5 F^2, giving rise to a NAND-memory density higher than the NOR one. The different cell size arises from the array organization and from the cell layout. In a NOR cell, a contact is shared every two cells, and basically this doubles the number of lithography features needed to define the cell contact. Moreover, the CHE programming does not allow an aggressive scaling of the cell-gate channel length, which instead occurs in NAND, where the cell-gate length and space define the technology node.

Although flash technology has clearly demonstrated its capability to shrink the cell size according to Moore's law, further reduction of the dimension is facing fundamental physical limits, and it is demanding technological developments that are making the cell scaling less convenient from the economic standpoint. As shown in Fig. 1, in the last two decades the reduction in flash-cell size has been achieved by simply scaling every dimension for both active (flash-cell transistor) and passive elements (interconnections). The technology enabler for such evolution has been the availability of advanced lithographic techniques based on the continuous reduction of the optical-source wavelength and on the reduction of the effective wavelength by interposing a suitable media between the lens and the silicon substrate. This latter technique, called immersion lithography, has been largely employed in recent years to enable the definition of lithographic features as small as 40–45-nm [7]. Although such equipment is really impressive from the technology and engineering standpoint, further techniques have been developed to shrink the cell size to the deca-nanometer range, while still relying on standard immersion lithography. All of them are based on the usage of sacrificial layers and deposition-etching-deposition techniques that enable doubling or even quadrupling the number of cells that can be defined inside the minimum lithographic pitch. Obviously these techniques come at the expense of much higher cost for cell production, thus creating a fundamental limitation to using them without a further improvement in the lithographic resolution. These additional developments in the lithographic equipment toward the extreme-UV (much shorter wavelengths) represent today the main technological challenge for enabling a further scaling of the planar flash-cell architecture at a reasonable production cost.

Apart from the technological issues for the further downscaling of flash-cell size, there are quite obvious physical limits that prevent further scaling of the cells. Such limitations have been successfully faced and managed over the years, but a few of them are seriously imposing strong compromises, in particular for cell reliability, in the sub-20 nm regime. The most obvious physical limitation is related to the dielectric thickness, namely the tunnel oxide dielectric and the interpoly dielectric. In particular, the minimum thickness of tunnel oxide is 9 nm for maintaining the same reliability specification, thus making the electrostatic control of the cell channel more and more difficult when scaling the transistor length.

Although the cell functionality has been preserved with scaling, cell-reliability degradation represents the main scaling issue in the most recent generations of flash technology. Such worsening is strictly related to the approach to a fundamental physical boundary: the number of electrons stored in the floating gate has been reduced from several thousands to hundreds or even tens of electrons, thus amplifying the impact of the dielectric degradation mechanisms. In fact, the impact of the usual trapping/detrapping mechanism, as well as trap-assisted tunneling phenomenon, is now much larger than in the past, and the loss of just few electrons from the floating gate is enough to delete the stored data, resulting in a significant reduction in overall reliability, particularly for the data-retention capability after cycling [9, 10].

Another side effect of the cell-size reduction is amplification of the noise effect, in particular the random telegraph-noise contribution. In fact, the usual 1/f noise of a MOS transistor has been observed as a giant threshold-voltage leap in sub-45 nm cell devices, thus adding an additional constraint to the overall reliability and noise immunity of this technology.

Finally, one of the most disturbing drawbacks associated with the cell-size reduction is the simultaneous reduction of the distance between adjacent cells. Such lower cell separation is responsible for additional issues in term of breakdown of the dielectric layer that separates adjacent word-lines and that need to maintain very high voltages. Moreover, the closer distance between adjacent cells makes the contribution of lateral capacitances between adjacent floating gates no longer

Fig. 3 Cell-proximity
interference effect in scaled
technologies. The
floating-gate potential of each
cell is capacitively coupled to
the potential of the adjacent
cells

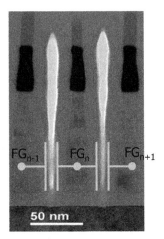

negligible, and the capacitive coupling used for determining the floating-gate
potential require taking this contribution into account.

In the standard, planar flash cell, the basic assumption is that the floating-gate
potential is controlled by capacitive-charge sharing where the main contributions
come from the control-gate terminal and from the cell-transistor nodes. However,
Fig. 3 shows that, in scaled flash technology, the planar-cell approximation is no
longer valid and that the short cell separation leads to additional capacitances that
must be considered in determining the floating-gate potential. Moreover, the lateral
capacitance contribution will depend on the potential of the adjacent floating gate,
thus linking the potential of each cell with the surrounding ones and leading to
so-called cell-proximity interference. In order to minimize the contribution of such
capacitances, a limited filling of the space between adjacent cells has been adopted,
thus leaving a void that introduces the smallest possible dielectric constant. Finally,
the parasitic capacitive coupling between neighboring floating gates leads to a low
coupling ratio with the control gate, which results also in a small stored charge
during the programming operation.

Finally, a third level of scaling challenges is related to the suitability of scaled
flash cells to meet the requirements of the new generation flash-based products.
NAND scaling is indeed increasing the density and lowering costs but reliability is
rapidly worsening, thus requiring more complex programming and reading algorithms for managing the higher bit-error-rate (BER). As result, write and read
latency is predicted to increase with the device scaling, thus opening the serious
question if the NAND-based SSDs price/performance ratio will be competitive with
magnetic disks.

Despite the prediction at the end of the last century that the floating-gate concept
faced severe technological limits beyond the 32-nm technology node,
NAND-memory density has shown the ability to be downscaled to the 16-nm node
with the multilevel-cell concept applied [11]. Such an achievement is partially due
to the fact that, for NAND products, a significant reliability drop (in particular for

endurance) is acceptable, considering the large ECC available and the processing power of dedicated controllers in SSD applications. A similar trade-off is not acceptable for NOR-flash applications, thus making their scaling more challenging. NOR flash has thus reached its scaling limitation at 45 nm (even considering the constant/declining market demand that is not fostering additional development efforts for this technology) while NAND is available at the 16-nm technology node.

2 The Future of the Floating-Gate Concept

The continuation of the flash scaling following Moore's law has been the main focus of the semiconductor memory industry during recent decades, and several strategies have been proposed throughout the years. These proposals can be schematically summarized into three main branches of developments: system-level management techniques, material engineering, and novel architectures.

2.1 System-Level Management Techniques

An effective workaround to manage the reduced reliability of scaled technology and the correspondingly higher bit-error-rate has been the massive introduction of Error-Correction-Code (ECC) algorithms. The object of the theory of error correction codes is the addition of redundant terms to the message, such that, on reading, it is possible to detect the errors and to recover the message that had most probably been written. Most popular ECC codes that correct more than one error are Reed-Solomon and BCH [12]. BCH and Reed-Solomon codes have a very similar structure, but BCH codes require fewer parity bits, and this is one of the reasons why they were preferred for an ECC embedded in the NAND memory [13]. While the encoding takes few cycles of latency, the decoding phase can require more cycles and visibly reduce read performance, as well as the memory response time at random access. Moreover, ECC requires additional cells to encode the logical data, thus using part of the memory chip just to improve the overall reliability.

Since the main scope of the device scaling is to reduce the cost per bit, a similar result can be obtained by exploiting the Multi-Level-Cell (MLC) capability of floating–gate devices. Planar NAND with MLC capability represents today the most cost-effective solution for an NVM concept. In fact, it can provide a minimum cell size of 4 F^2 combined with MLC capabilities up to 3 bits-per-cell at the leading-edge technology node of 16 nm, resulting in an effective cell size of 1.3 F^2. MLC feasibility is mainly related to the system-level management, and it require a more precise level placement (smarter programming algorithms), as well as the capability to manage worse BER and reliability (a high level of ECC is required).

The obvious advantage of a 2 bit/cell implementation (MLC) with respect to a 1 bit/cell device (SLC) is that the area occupied by the matrix is half as large; on the other hand, the area of the periphery circuits, both analog and digital, increases. This is mainly due to the fact that the multilevel approach requires higher voltages for programming (and therefore bigger charge pumps), higher precision and better performance in the generation of both the analog signals and the timings, and an increase in the complexity of the algorithms. Driven by cost, flash manufacturers are now developing 3 bit/cell (eight threshold voltage levels) and 4 bit/cell (16 levels). Three- and four-bits per cell are usually referred to as XLC (8 and 16 LC, respectively).

The capability to mitigate the scaling issues with the adoption of system-level management techniques is well represented by the adoption of a sophisticated memory controller to manage the NAND flash inside products. The original aim of the memory controller was to provide the most suitable interface and protocol suitable to both the host and the flash memories. However, with the intrinsic reliability and performance degradations observed in sub-45 nm flash devices, the role of the memory controller has been extended, and now it is employed to efficiently handle data, maximizing transfer speed, data integrity, and information retention.

For example, in a typical consumer application, not all the information stored within the same memory location changes at the same frequency, and some data are often updated while others remain the same for a very long time. It's clear that the cells containing frequently-updated information are stressed with the large number of write/erase cycles, while the cells containing information updated very rarely are much less stressed. In order to mitigate the reliability issues, it is important to keep the aging of each block of cells at a minimum and as uniform as possible, and to monitor the maximum number of allowed program/erase cycles. To this aim, wear leveling techniques are employed to dynamically map the logical data onto different cells, keeping track of the mapping. In this way, all the physical cells are evenly used, thus keeping the aging under a reasonable value.

2.2 Dielectric-Materials Engineering

The second scaling strategy explored during the last 15 years has been the improvement of flash-cell active materials, in particular the dielectrics employed in the floating-gate definition. One of the most important attempts made to mitigate scaling limitations, while retaining the very high integration density of NAND-flash architecture, has been the attempt to replace the conventional floating gate with a charge-trapping layer. Silicon nano-crystal trapping layers have been investigated in the past [14], but they present a few drawbacks, like reduced threshold shift and the presence of percolation paths between source and drain that become more severe with the scaling of the cell size. The silicon nano-crystal technology requires a careful control of the nano-dots size, dimension, shape, and density, because these parameters significantly impact device performance and reliability. Moreover the down-scaling of this technology was expected to be difficult beyond the 32-nm technology node due to the minimum nano-crystal size that has so far been achieved in a reproducible way.

Other alternatives include the use of a continuous trapping layer (charge-trap memories, also called CT memories), like silicon nitride in the SONOS-device architecture [15]. This approach promises to solve several of the scalability issues: the charge is trapped in a thin dielectric layer, and therefore there is no problem of capacitive interference between neighboring cells; since the charge is stored in electrically insulated traps, the device is also immune to SILC, the parasitic leakage current caused by single defects in the dielectric layer, while, in conventional floating-gate devices, even a single defect can discharge the whole floating gate, which is a conductive storage medium; the replacement of the floating gate with a trapping layer reduces the overall thickness of the gate stack, and allows for easier integration of the cell in the CMOS process. Even if alternatives are available, silicon nitride is probably the best storage material since it is characterized by a high trap density and by a very long lifetime of the charged state that ensures large threshold windows and excellent data retention in memory applications [16].

In this architecture, the charge is trapped in a silicon nitride layer, inserted between two silicon-oxide layers, which act as a tunnel dielectric and blocking layer to prevent charge injection from the control gate. Although this cell architecture is known since the 1980s and in spite of its better compatibility with standard CMOS process flow and lower costs, it lost ground in favor of the floating-gate one, because of several fundamental problems: first, cell programming is limited by the erase saturation, which takes places because of the parasitic electron injection from the control gate through the top oxide, balancing the hole injection from the sub-strate. Second, he thinning of the tunnel oxide (<2.5 nm) improves the threshold window and the programming speed, but results in poor retention, even at room temperature, because of direct tunnelling through the tunnel oxide, and charge mobility in the nitride layer. Finally, increasing the tunnel thickness improves the retention, but requires larger programming voltages; it also reduces the speed and activates the tunnelling through the top oxide.

Despite the scaling issue of the floating-gate concept, the actual drawbacks of the CT memories made conventional planar flash a preferred choice for mainstream technologies, thus limiting the usage of CT layers in planar flash production.

2.3   Novel Flash Architectures

In this scenario and considering the growing difficulties that must be faced for planar NAND scaling, there is a strong interest in the so-called 3D NVM tech-nologies. This is the third scaling strategy employed by the semiconductor memory industry, where novel cell architectures are proposed to further extend the scaling. Among the possible architectures, the most promising are the vertical ones, i.e., all the architectures that try to exploit the vertical direction to increase the cell density,

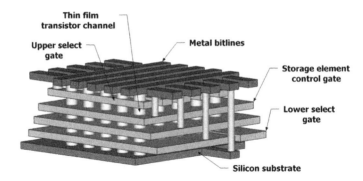

Fig. 4 Schematic representation of a vertical 3D-NAND architecture

thus moving from the conventional planar cell construction to a three dimensional (3D) approach.

There are currently two main trends in the efforts to develop 3D structures that can provide a higher integration density: cross-point memory array, where several memory layers can be stacked, and the so called 3D NAND, where the standard NAND strings are integrated along the vertical dimension (Fig. 4).

3D NAND was proposed as a cost-effective solution for vertical NAND fabrication. The process is intended to use a minimum number of masks, thus reducing the fabrication costs. However much effort is required to develop the suitable integration modules in order to have at least 32 layers along the vertical dimension. Such a huge number of layers is needed to make it possible for a 3D NAND with a relaxed pitch to be cost-effective with respect to the existing planar NAND.

Although several approaches have been proposed to fabricate a 3D NAND chip, the most effective solution employs a vertical channel with a horizontal gate. In this case, the number of critical masks is low, since the entire stack is etched at the same time. The limited dependence of wafer cost on the number of levels results in a fortunate economic scenario for these arrays, even if the typical cell size is relatively large and many stacked layers are necessary to reach a small equivalent cell area (i.e., a single cell area divided by the number of memory layers). The typical 3D NAND structure is illustrated in Fig. 4, the quantity of cells inside a string is defined by the number of vertical wordlines layers stacked in the array. Bitlines and drain selector lines run horizontally and are used to select string in the two directions. Three architectures with vertical channels and horizontal gates are: BiCS [20], VRAT [21], and TCAT [22]. It has been announced that the 3D-NAND technology will be commercialized starting with the SSD application, with 24 layers and MLC in a 128-Gb monolithic module based on CT technology [18] and with 32 layers 256-Gb MLC and 384-Gb triple-level cell (TLC) 3D NAND based on floating-gate technology [19].

3 Alternative Storage Concepts

To improve the performance and scalability with respect to floating-gate devices, innovative concepts for alternative NVM have been proposed in the past and are under investigation today, as we dream of find the ideal memory that combines fast read, fast write, non-volatility, low-power, and unlimited endurance, and obviously at a cost comparable to flash or DRAM.

Table 1 reports a schematic grouping of the alternative NVM concepts based on the decoding technique and on the selected architecture. Generally speaking, electronic memories can be divided into two main classes: solid-state-device memory, where each cell is placed at the intersection of two orthogonal metal lines defined through lithographic technique (e.g., DRAM, Flash); and mechanically decoded memories in which a mechanical positioning of a programming/reading equipment is adopted to address data on a flat substrate (e.g., magnetic disc and optical disc).

In 2000, IBM proposed an alternative memory device [23], basically a sort of miniaturized hard disk system based on a micro-electro-mechanical system (MEMS) that actuates thousands of tips capable of decoding the information stored on a flat media. Several media were proposed at that time to store data with a bit size in the range of ten nanometers, but the overall complexity and cost of this system did not allow it to compete with the fast-growing NAND technology.

On the other side, the more traditional electronically decoded memories tried to exploit the properties of novel materials to create self-selecting (cross-point) and transistor-selected NVM memories. The main class of emerging NVM technologies so far investigated is based on inorganic materials, and it includes the alternative memory concept with the highest maturity level, namely Ferroelectric Memories

Table 1 Schematic grouping of the alternative NVM concepts

Electronic decoded, lithography depend (Moore's law follower)	Mechanical decoded, lithography independent (beyond Moore's law)
• Transistor selected (like DRAM or Flash) – Ferroelectric memory (FERAM) – Magnetoresistive memory (MRAM and STT-MRAM) – Resistive RAM (RRAM) – Phase-Change Memory (PCM)	• Probe storage (Seek and scan, like Hard Disk or CD) – Polymers – Chalcogenide – Ferroelectric
• Cross-point memories (Passive arrays) – Ferroelectric polymers (PFRAM or TFEM) – Organic charge-transfer complex (conductive polymers) – Resistive switching	

(FeRAM), Magnetoresistive Memories (MRAM), Phase Change Memories (PCM) and Resistive RAM (ReRAM). These NVM alternative concepts will be described in detail in the next sections.

3.1 Ferroelectric Memories

FeRAM is one of the few alternative NVM that has been commercialized so far, even if at a technology node much more relaxed than the one used for flash memories and with several challenging technological problems, mainly related to new materials and new manufacturing technologies. Two classes of ferroelectric materials are currently used for FeRAM memories: perovskite structures and layered structures. Actually, the most-used perovskite material for ferroelectric memories is $PbZr_xTi_{1-x}O_3$, also called PZT, while the layered ferroelectric choices for FeRAM memories are either strontium-bismuth-tantalate $Sr_{1-y}Bi_{2+x}Ta_2O_9$, also called SBT, or lanthanum substituted-bismuth-titanate $Bi_{4-x}La_xTi_3O_{12}$, also called BLT. Among them, the commercially preferred option is represented by the PZT, usually deposited with the MOCVD technique at temperatures higher than 600 °C. Ferroelectric materials can be polarized spontaneously by an electric field. The polarization occurs as a lattice deformation of the cubic form below the Curie point, the temperature above which the material becomes paraelectric. For example, in PZT the titanium atom can be moved by an electric field to two stable positions that are above and below the oxygen plane of the structure. An important property of ferroelectric materials is therefore their residual permanent polarization, typically in the range of 10–30 $\mu C/cm^2$. The voltage required to switch the permanent polarization is in the range of 1.5–3 V for typical deposited-layer thicknesses of 70–100 nm. It follows that ferroelectric memories can be a valuable solution for low-power and very low-voltage application, like a battery-operated embedded system, smartcards, and RFID applications.

One of the most challenging features of this technology is presented by the integration of the ferroelectric layer into the standard CMOS process. The bottom and top electrodes of the ferroelectric layer must be realized with a specific alloys usually constituted by iridium and platinum. Hydrogen contamination of the ferroelectric material must be avoided in order to prevent the reduction of the permanent-polarization capability, thus requiring a specific barrier layer all around the ferroelectric capacitor. At the same time, oxygen can diffuse during the high-temperature treatment required for ferroelectric alloy deposition, oxidizing the underlying metal layers. Finally, special care must be devoted to the definition of the capacitor shape through a dry etching, to the final dielectric layer used to seal the capacitor from the surrounding environment and to the effect of the plasma damage due to the CMOS back-end process, resulting in a possible discharge through the capacitor that destroys its capability to store data.

Endurance, also called electric fatigue, is an important reliability characteristic, and it is related to the decrease of the ability to switch the memory cell into the

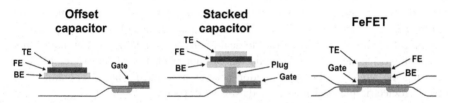

Fig. 5 FeRAM cell architectures with the ferroelectric material (FE) integrated either into a separate storage element, i.e., a ferroelectric capacitor (offset capacitor or stacked capacitor option) or into the selection element, i.e., a ferroelectric FET transistor (FeFET)

opposite state, after being kept programmed in one state for long periods of time. This effect is related to the polarization shift in the hysteresis loop, and it is proportional to the increasing number of switching cycles. Nevertheless, the write cycle (but also the read cycle, since several proposed cell structures have a destructive reading) is expected to have an endurance level of about 10^{12}, which will be enough for a wide majority of high-demanding storage applications. Up to now several FeRAM cell structures have been proposed, with the ferroelectric material integrated either into a separate storage element, i.e., a ferroelectric capacitor [24], or into the selection element, i.e., a ferroelectric FET transistor [25] (Fig. 5). In the latter case, the storage and the selection elements are merged. The first cell type can be used in both the two-transistor/two-capacitor (2T/2C) cell and the one-transistor/one-capacitor (1T/1C) cell, while the latter has been proposed with a one-transistor (1T) approach. Moreover, a NAND-type FeRAM array configuration has also been proposed with the name of chain-type FeRAM memory. All FeRAM architectures have high-speed access cycles and provide genuine random access to all memory locations. Among the proposed architectures, the 1T/1C FeRAM approach is characterized by quite large cell size that cannot compete with today's high-density solutions (flash for NVM and DRAM for volatile storage). However, the very low-voltage operation and the superior electrical performance in terms of programming speed and endurance make FeRAM technology a valuable solution for specific application in the embedded market. The 1T FeRAM architecture is a very promising alternative for high-density application with very small cell size and practically an infinite endurance level, but its processing has proven to be very complicated, and no products have been so far announced or released. Apart from the FeRAM potentialities discussed above, there is a quite important issue for cell scaling that could impact the further development of this technology. In fact, the cell sensing in capacitor-based architectures relies on the capability to detect the displacement current associated with this capacitance, similarly to what is usually done in DRAM technology. With a planar capacitor approach, the continuous shrinking of the cell size corresponds to a reduction of the capacitor surface, degrading the available signal for reading the cell status. As for DRAM, this issue can be solved just by moving from the simple planar capacitor to the more complicated three-dimensional (3D) capacitor architecture, analogously with what had already happened in the DRAM-scaling roadmap. Moreover, ferroelectric properties tend to disappear in very thin layers, thus making the scaling of the active material below 50-nm thickness a huge issue, at least for the proposed ferroelectric alloys. It follows that, to scale FeRAM technology below the 90-nm technological node, a more complicated 3D approach is mandatory, really challenging the already difficult fabrication process associated with the integration of the ferroelectric material into a standard CMOS process.

3.2   Magneto-Resistive Memories

Up to the early 2000s, all the development efforts for MRAM technology were MTJ cell based [26], with an architecture composed of one transistor and one resistor (1T/1R). This technology relies on the adoption of a tunnel junction coupled to magneto-resistive materials that exhibit changes in their electric resistance when a magnetic field is applied. The MTJ is composed of a pinned magnetic layer, a tunnel barrier, and a free magnetic layer. Electrons spin polarized by the magnetic layers traverse the tunnel barrier. A parallel alignment of the free layer with respect to the pinned layer results in a low resistance state, while an anti-parallel alignment results in a high resistance state [27, 28]. Therefore the storing mechanism consists of the permanent magnetization of the ferromagnetic material in the MTJ. The datum can be sensed as the resistance in the MTJ which can be high (low current) or low (high current). The writing can be performed through the magnetic field produced by the current flowing in the bit- and digit-lines.

The non-destructive read with a very fast access cycle is the premise for high performance, equal-long read and write cycles, and for low-power operation. Moreover, the structure is radiation-hard with a potentially unlimited read/write endurance, which makes MRAMs suitable for write intensive storage applications. The major MRAM disadvantage appears to be the high write current. While this technology has enough read current to guarantee a fast access time, it requires a very large write current (mA range), which increases power consumption. The current requirements become even more challenging when MRAM devices are scaled. In fact, since the data-retention capabilities of MRAM memory cells are related to the total volume of magnetic material used in the free layer of the MTJ, it is expected that the capability to retain the stored information will be degraded in scaled devices, where the MTJ geometrical features are shrunk. It follows that suitable materials with higher magnetic coercivity must be adopted to retain the data in scaled devices, thus demanding more current to be programmed and erased.

One of the key milestones for the development of MRAM technology was the introduction of the toggle-MRAM writing scheme in order to achieve better program-disturb immunity. With respect to the conventional MRAM, a toggle-MRAM employs a programming technique based on the current amplitude (as in conventional MRAM) and on the timing of the applied programming pulses. Only the correct sequence of pulses delivered to the selected cell are able to switch its magnetization, leaving unchanged all the other cells along the programmed bitline and wordline. Beginning in 2006, low-density (4- to 16-Mb) chips based on the toggle MRAM concept have been commercialized and are today available on the market for very specific applications.

In order to mitigate the scaling issues of the MRAM concept, a novel programming technique has been recently investigated and is fuelling a renewed interest in the MRAM technology. This approach is based on the spin-polarizing effect [29], in which magnetization orientations in magnetic multilayer nanostructures can be manipulated via spin-polarized current. The so-called Spin-Transfer Torque-MRAM (STT-MRAM) technology is based on an MTJ structure where a current-induced switching caused by spin-transfer torque is exploited. Despite that this approach enables mitigation of some of the conventional MRAM issues, particularly for scaling, there are still several challenges that must be faced (e.g., self-read disturbance, writing times, cell integration). Today the STT-MRAM developments are very active with prototypes of 64 Mb and 1 Gb at 90 nm [30] and 54 nm [31], respectively. However no volume production has yet been started.

3.3 Phase-Change Memories

Among the different NVM based on mechanisms alternative to the floating-gate concept, Phase-Change Memories (PCM) are one of the most promising candidates to become mainstream NVM, having the potentiality to improve the performance compared to flash — random access time, read throughput, direct write, bit granularity, endurance — as well as to be scalable beyond flash technology.

PCM exploits thermally reversible phase transitions of some chalcogenide materials or alloys (e.g., $Ge_2Sb_2Te_5$). In fact some alloys based on the VI group elements (usually referred to as chalcogenides) have the interesting characteristic of being stable at room temperature, both in their amorphous and crystalline phases. In particular, the most promising are the GeSbTe alloys that follow a pseudo-binary composition (between GeTe and Sb_2Te_3), hereafter referred to as GST. The most interesting feature of these alloys is their capability to reversibly switch between a high-resistance amorphous phase and a low-resistance crystalline one in a few hundreds of nanoseconds.

The basic cell structure is composed of one transistor and one resistor (1T/1R) that can be programmed through the current induced Joule heating, and can be read by sensing the resistance change between the amorphous and the polycrystalline phase. The PCM cell is essentially a resistor of a thin-film chalcogenide material with a low-field resistance that changes by orders of magnitudes, depending on the phase state of the GST in the active region. The switch between the two states occurs by means of local temperature increases. Above the critical temperature, crystal nucleation and growth occur, and the material becomes crystalline (Set operation). To bring the chalcogenide alloy back to the amorphous state (Reset operation), the temperature must be increased above the melting point of hundreds of °C and then very quickly quenched down to preserve the disorder and not let the

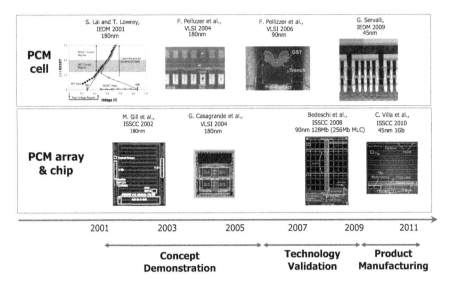

Fig. 6 Evolution of PCM development

material crystallize. From an electrical point of view, it is possible to use the Joule effect to reach locally both critical temperatures using the current flow through the material by setting proper voltage pulses. The cell read out is performed at low bias. Programming thus requires a relatively large current in order to heat-up the GST and results in to a thermally induced, local phase change. Phase-transitions can be thus easily achieved by applying voltage pulses with various amplitudes and with durations in the range of tens or hundreds of nanoseconds.

Although the phase-change concept is well known for years and the first studies date back to the 1970s [32, 33], its application for NVM experienced renewed interest in 1999, when the idea for integrating a phase-change material into a NVM cell was presented again [34]. As result, the effort to bring the PCM basic concept to a mature technology level has constantly increased since then, and many groups have started to study, to develop, and to integrate a Multi-Megabit array in the memory cell.

The development roadmap of PCM technology is summarized in Fig. 6. A 180-nm technology node has been used to develop the first demonstrator vehicles and to prove the technology viability [35]. The BJT-selected cell has been chosen for the high-performance and high-density application, since the cell size can be ~ 5 F^2. The MOS-selected cell is suitable for system-on-a–chip or embedded application [36], because, in spite of the larger cell size (~ 20 F^2), the memory integration adds only very few masks to the logic process with a clear cost advantage.

A 90-nm technology node has been developed and commercialized using a 128-Mb product [37]. A 1-Gb-PCM product fabricated at the 45 nm technology node with a cell size of 5.5 F^2 has been developed, and it is in volume production for wireless applications [38]. This 45-nm-PCM architecture [39] demonstrates the maturity of the technology. The energy delivered to program a bit is on the order of 10 pJ, with a state-of-the-art, random access time of 85 ns, read throughput 266 MB/s and write throughput 9 MB/s [40]. These peculiar features combined with data retention, single bit alterability, execution in place, and good cycling performance enable traditional NVM utilizations but also novel applications in the LPDDR field. Moreover, PCM is considered the essential ingredient to push to the market the so called Storage-Class Memory (SCM) [41], a non-volatile, solid-state-memory technology that is capable of filling the gap between CPUs and disks.

One of the key challenges of the PCM-cell scaling is the reduction of the power, mainly due to the programming current. So far, most of the development has been focused on the cell-structure optimization in order to increase the programming efficiency. More recently it has been shown that, also by proper engineering of the active material, it is possible to reduce the programming current of a standard PCM cell of 1/20 [42], enabling even more low-power, innovative applications.

3.4 Resistive RAM

Resistive Memories, usually called ReRAM, are a quite large class of memory concepts that store the information in the resistance value of the cell. It is worth noting that PCM and MRAM fall within this definition too, but the interest and the efforts devoted to their specific development in the last decades set them apart as stand-alone concepts. All the resistive memories are electrically programmable, although they are based on different material classes and different proposed switching mechanisms:

- Formation of metallic bridges by dielectric breakdown and bridge opening by thermal fusion [43];
- "Volume" switching, e.g., by electronic charge transfer (redox) mechanisms [44];
- Switching of filament regions in the resistive material by electronic charge transfer (redox) mechanisms [45];
- Electrochemical growth and dissolution of metallic dendrites by solid electrolyte/electrode processes (programmable metallization cells, PMC) [46].

The resistive memory materials proposed in these concepts range from organic materials (rotaxanes and catananes, polyphenyleneethylenes, Cu- and Ag-TCNQ) to inorganic (chalcogenide alloys, perovskite-type oxides, manganites, binary transition metal oxides), while electrode materials comprise various metals as well as electronically conducting oxides and nitrides. Among the materials that exhibit a resistive switching phenomenon oxide materials have been studied intensively.

On the basis of *I–V* characteristics, the switching behaviors can be classified into two types: unipolar (nonpolar) and bipolar. In unipolar-resistive switching, the switching direction depends on the amplitude of the applied voltage but not on the polarity. An as-prepared memory cell is in a highly resistive state and is put into a low-resistance state (LRS) by applying a high-voltage stress. This is called the 'forming process'. After the forming process, the cell in a LRS is switched to a high-resistance state (HRS) by applying a threshold voltage ('reset process'). Switching from a HRS to a LRS ('set process') is achieved by applying a threshold voltage that is larger than the reset voltage. In the set process, the current is limited by the current compliance of the control system or, more practically, by adding a series resistor. This type of switching behavior has been observed in many highly insulating oxides, such as binary metal oxides.

Bipolar resistive switching shows a resistive change that depends on the polarity of the applied voltage. This type of resistive switching behavior occurs in many semiconducting oxides, such as complex perovskite oxides.

It should be noted that there is a lot of speculation and controversy on the actual, physical switching mechanisms for many of these concepts. Moreover, in many cases, the role of the electrode materials is found to be very important although it is also not exactly understood. Independent of the mechanism, however, the important, basic characteristics for all concepts are represented by the required switching voltage and switching current. Indeed, low switching voltages are required to be compatible with the low supply voltages of scaled technologies, while low switching currents are required to be able to switch with minimal-size selector devices, as well as to limit the switching power. For array fabrication, a transistor-type architecture is preferred while the cross-point architecture and the diode architecture open the path toward stacking memory layers, and therefore are ideally suited for mass-storage devices.

An important advantage of ReRAM technology is the good compatibility with the CMOS processes, in particular for binary-oxide-based memories. The critical issues for the future development of ReRAM devices are reliability, such as data retention and memory endurance (the number of erase and program cycles), and the characteristic variations from cell to cell and from chip to chip. Despite the large potential evident in some of these concepts, the poor control of resistance distribution and the low maturity level reached by the cell integration into sub-micrometric devices represent today the main limitations for several of them.

Despite the huge interest among the semiconductor companies and inside the scientific community, very few attempts to develop ReRAM have materialized as real commercial products. The availability of an evaluation kit including a ReRAM memory fabricated in 180-nn technology has been announced [47]. On the research level, a 32-Gb, ReRAM test chip developed in a 24-nm process, with a diode as the selection device and a 2-layered architecture, has been presented [48]. These interesting results make ReRAM one of the most promising alternative memory technologies for mainstream applications.

4 Scaling Path and Issues in Various Emerging Architectures

The current NVM mainstream is based on flash technology, and it is expected that flash will be the high-volume NVM in production for the next years. Flash technology is characterized by a compact structure in which the selecting element and the storage element are merged in a MOS-like architecture. The resulting, full compatibility with the CMOS technology and the compact device size have made flash technology the cheapest solution for stand-alone and embedded-memory applications. However, discordant requirements to shrink the MOS structure, while preserving good selection and storage capabilities, are making flash scaling more and more difficult. Moreover, even if in the long term the cost advantage is important, better performance can speed up a novel technology introduction. In fact, a NVM with low-power and low-voltage capabilities, bit granularity, fast operations, and higher endurance would be a potential game changer for system designers.

To enlarge application segments, offering better performance and scalability, new materials, and alternative memory concepts are mandatory to boost the NVM industry. During recent decades, a total of more than 30 NVM technologies and technology variations have been competing for a piece of the fast growing NVM market, many of them aiming to replace also DRAMs. Although the planar-NAND scaling is becoming harder and it is obtained with several compromises in term of reliability, state-of-the-art NAND technology is 3 bits-per-cell 16 nm. Moreover, 3D NAND is becoming a mature technology. It follows that any NVM development must be able to provide the same capabilities at higher density. If this is not achieved in the next generation, technology scaling will not result in a cost reduction, thus eliminating the interest to continue along this path. Cost structure is therefore a fundamental parameter for benchmarking novel NVM concepts, in particular for those that want to compete for data-storage solutions, where MLC capabilities and/or 3D-stacking are mandatory.

On the other hand, the actual development efforts of alternative NVM concepts are demonstrating that disruptive innovation takes a long time. Figure 6 recounts schematically the development history of PCM technology, the only emerging memory concept that has reached the volume production maturity for large-density arrays. It is worth noting that the continuous need to stay close to the state-of-the-art lithographic node, combined with the necessary learning cycles, necessitated a decade of effort to evolve from concept to mass production. Moreover, we need also to consider the time-to-market, i.e., the time needed to get a significant profit from a novel NVM technology. For the flash NOR, about four years were needed to reach the break-even with the EEPROM in terms of profit for the main semiconductor industries involved in this market. The scaling lesson was also clearly demonstrated by the MRAM developments. MRAM products reached product maturity simultaneously with the last feasible technology node for the conventional toggle-MRAM, thus limiting the commercial success of MRAM technology and reducing it to a niche market. Only the discovery and exploitation of the STT-MRAM concept enabled a better scalability below the 90 nm node, thus renewing the interest in this technology.

In this perspective, there are two major aspects that must be considered for evaluating the potentials of any emerging memory concept, namely the readiness for moving beyond the leading—edge technology node and the scalability perspective. If we combine the above two statements, it follows that any realistic proposal for a novel NVM technology must prove its feasibility for the sub-1X-nm-technology node.

References

1. R. Bez et al., "Introduction to Flash Memory", Proceedings of the IEEE, vol. 91, n. 4, 2003.
2. "Flash Memories", edited by P. Cappelletti, C. Golla, P. Olivo, E. Zanoni, Kluwer Academic Publishers, 1999.
3. G. Ginami et al., "Survey on Flash Technology with Specific Attention to the Critical Process Parameters Related to Manufacturing", Proceedings of the IEEE, vol. 91, n. 4, p. 503, 2003.
4. A. Fazio, "A High Density High Performance 180 nm Generation ETOX Flash memory Technology", IEEE IEDM Tech. Dig. pp. 267–270, 1999.
5. S. Keeney, "A 130 nm Generation High Density ETOX Flash Memory Technology", IEEE IEDM Tech. Dig. pp. 2.5.1–2.5.4, 2001.
6. G. Servalli et al., "A 65 nm NOR Flash Technology with 0.042 μm^2 Cell Size for High Performance Multilevel Application", IEEE IEDM Tech. Dig., pp. 2.5.1–2.5.4, 2005.
7. H. Hu et al., "K = 0.266 immersion lithography patterning and its challenge for NAND FLASH", Semiconductor Technology International Conference (CSTIC), 2015 China.
8. K. Naruke et al., "Stress Induced Leakage Current Limiting to Scale Down EEPROM Tunnel Oxide Thickness", IEEE IEDM Tec. Dig., pp. 424–427, 1988.
9. J. S. Witters et al., "Degradation of Tunnel Oxide Floating Gate EEPROM Devices and Correlation with High Field Current Induced Degradation of Thin Gate Oxide", IEEE Trans. Electron Devices, vol. 36, p. 1663–1682, 1989.
10. D. Ielmini et al., "A Statistical Model for SILC in Flash Memories", IEEE Trans. Electron Devices, vol. 49, n. 11, p. 1955–1961, 2002.
11. Micron Press Release, "Intel, Micron Extend NAND Flash Technology Leadership With Introduction of World's First 128 Gb NAND Device and Mass Production of 64 Gb 20 nm NAND", December 6, 2011.
12. R. Micheloni et al., "Error Correction Codes for Non-Volatile Memories", Springer-Verlag, 2008.
13. R. Micheloni et al., "A 4 Gb 2b/cell NAND Flash Memory with Embedded 5b BCH ECC for 36 MB/s System Read Throughput", IEEE International Solid-State Circuits Conference Dig. Tech. Papers, pp. 142–143, Feb. 2006.
14. B. DeSalvo et al., "How Far Will Silicon Nanocrystals Push the Scaling Limits of NVMs Technologies?" IEEE IEDM Tech. Dig., p. 597–600, 2003.
15. Y. Shin et al., "A Novel NAND-type MONOS Memory using 63 nm Process Technology for Multi-Gigabit Flash EEPROMs", IEEE IEDM Tech. Dig., p. 327–330, 2005.
16. B. Eitan et al. "NROM: A Novel Localized Trapping, 2-bit Nonvolatile Memory Cell", IEEE EDL, Vol. 21, No. 11, 2000.
17. C. H. Lee et al., "A novel SONOS structure of SiO2/SiN/Al2O3 with TaN metal gate for multi-Giga bit Flash memories", IEDM tech. digest 2003.

18. J. Kim et al., "Novel Vertical-Stacked-Array-Transistor (VSAT) for ultra-high-density and cost-effective NAND Flash memory devices and SSD (Solid State Drive)", Symposium on VLSI technology 2009.

19. Micron Press Release, "Micron and Intel Unveil New 3D NAND Flash Memory", March 26, 2015.

20. H. Tanaka et al., "Bit Cost Scalable technology with punch and plug process for ultra-high density Flash memory", Symposium on VLSI technology 2007.

21. J. Kim et al., "Novel 3-D structure for ultra-high density Flash memory with VRAT (Vertical-Recess-Array-Transistor) and PIPE (Planarized Integration on the same Plane)", Symposium on VLSI technology 2008.

22. J. Jang et al., "Vertical cell array using TCAT (Terabit Cell Array Transistor) technology for ultra-high density NAND Flash memory", Symposium on VLSI technology 2009.

23. P. Vettiger et al., "The "Millipede"-More than thousand tips for future AFM storage", IBM Journal of Research and Development, vol. 44, n. 3, pp. 323–340, 2000.

24. S.-H. Oh et al., "Novel FERAM Technologies with MTP Cell Structure and BLT Ferroelectric Capacitors", IEEE IEDM Tech. Dig., p. 835–839, 2003.

25. H. Ishiwara, "Recent Progress in FET-Type Ferroelectric Memories", IEEE IEDM Tech. Dig., p. 263–267, 2003.

26. M. Durlam et al., "A 0.18 um 4 Mb Toggling MRAM", IEEE IEDM Tech. Dig., p. 995–999, 2003.

27. S. Tehrani et al., "Magnetoresistive Random Access Memory Using Magnetic Tunnel Junctions", Proceedings of the IEEE, vol. 91, n. 5, p. 703–714, 2003.

28. J. C. Slonczewski, "Current-Driven Excitation of Magnetic Multilayers", Journal of Magnetism and Magnetic Materials, vol. 159, n. 1–2, p. L1–L7, 1996.

29. C. Demerjian, "Everspin Makes ST-MRAM a Reality", "LSI AIS 2012: Non-volatile Memory with DDR3 Speeds", SemiAccurate.com, November 16, 2012.

30. S. Chung et al., "Fully Integrated 54 nm STT-RAM with the Smallest Bit Cell Dimension for High Density Memory Application", IEEE IEDM Tech. Dig., p. 12.7.1–12.7.4, 2010.

31. S. R. Ovshinsky, "Reversible Electrical Switching Phenomena in Disordered Structures", Phys. Rev. Lett., vol. 21, p. 1450, 1968.

32. R. G. Neale et al., "Nonvolatile and Reprogrammable, the Read-Mostly Memory is Here", Electronics, p. 56, Sept., 1970.

33. G. Wicker, "Nonvolatile, High Density, High Performance Phase Change Memory" SPIE Conf. on Elect. and Struc. for MEMS, Australia, 1999.

34. F. Pellizzer et al., "Novel µTrench Phase-Change Memory Cell for Embedded and Stand-Alone Non-Volatile Memory Applications", Symp. on VLSI Tech., p. 18–19, 2004.

35. F. Ottogalli et al., "Phase-Change Memory Technology for Embedded Applications", Proc. ESSDERC 04, p. 293–296, 2004.

36. F. Bedeschi et al., "A Multi-Level-Cell Bipolar-Selected Phase-Change Memory", Solid-State Circuits Conference, ISSCC, p. 428, 2008.

37. G. Servalli, "A 45 nm generation Phase Change Memory technology", IEEE IEDM Tech. Dig., p. 1–4, 2009.

38. Micron Press Release, "Micron Announces Availability of Phase Change Memory for Mobile Devices", July 18, 2012.

39. C. Villa, D. Mills, G. Barkley, H. Giduturi, S. Schippers, D. Vimercati, "A 45 nm 1 Gb 1.8 V Phase-Change Memory", Solid-State Circuits Conference, ISSCC, p. 270–271, 2010.

40. R. F. Freitas and W. W. Wilcke, "Storage-Class Memory: The next Storage System Technology", IBM Journal of Research and Development, vol. 52(4/5), p. 439–448, 2008.

41. T. Shintami et al., "Properties of Low-Power Phase-Change Device with GeTe/Sb2Te3 Superlattice Material", EPCOS 2011, p. 110, 2011.

42. K. Szot et al., "Localized Metallic Conductivity and Self-Healing during Thermal Reduction of SrTiO3", Phys. Rev. Lett., vol. 88, n. 7, p. 075508, 2002.

43. T. Iizuka-Sakano et al., "Stability of the Staging Structure of Charge-Transfer Complexes Showing a Neutral–Ionic Transition", Phys. Rev. B, vol. 70, n. 8, p. 085111, 2004.

44. B. J. Choi et al., "Resistive Switching Mechanism of TiO2 Thin Films Grown by Atomic-Layer Deposition", J. Appl. Phys., vol. 98, n. 3, p. 33715, 2005.
45. M. N. Kozicki et al., "Nonvolatile Memory Based on Solid Electrolytes", NVMTS 2004.
46. Panasonic Press Release, "The New Microcontrollers with On-Chip Non-Volatile Memory ReRAM", May 15, 2012.
47. T.-Y. Liu et al., "A 130.7mm2 2-Layer 32 Gb ReRAM Memory Device in 24 nm Technology", Solid-State Circuits Conference, ISSCC, pp. 210, 2012.

Performance Demands for Future NVM

Roberto Gastaldi

1 Introduction

Since their introduction around 25 years ago, flash memories have known an immediate fortune, becoming the mainstream Non-volatile memory technology. The first to be used was NOR flash, which is based on hot-electrons for programming and Fowler-Nordheim tunneling for erasing, followed after some years by NAND architecture based on tunneling for both programming and erasing.

Flash NOR architecture is capable of fast random reading, but write power consumption is high because of the hot-electrons mechanism and writing time is not very fast (on the order of tens of μs) leading to limited write throughput. In addition, the cell is quite large, on the order of 10 F^2. To its 5 F^2 cell and Fowler-Nordheim tunneling used to write and erase, the NAND approach can have a much more compact array and less power consumption during write and erase operations. This opens the possibility to reading/writing an entire kbyte-size page in parallel, managing the data output in burst mode. This approach makes possible overcoming the high latency for random reading typical of NAND and achieving very high write-throughput, and this resulted in NAND-Flash establishment in the mass memory market and the replacement of NOR, spanning the scaling path down to sub-20-nm technologies. Today, however, NAND development is encountering increasing difficulties, in part, due to physical limits of the technology, and in part due a new performance demands from the market. Memory technologies based on completely new concepts have been in development for many years, and some of them have the possibility to replace NAND in the future [1].

R. Gastaldi (✉)
Redcat Devices s.r.l, Milano, Italy
e-mail: r.gastaldi@redcatdevices.it

© Springer International Publishing AG 2017
R. Gastaldi and G. Campardo (eds.), *In Search of the Next Memory*,
DOI 10.1007/978-3-319-47724-4_4

2   Evolution of NAND

NAND is effective in permanently storing large quantities of data but its cycling have been degraded during the NAND scaling roadmap for physical reasons, and it is limited to few thousands of cycles in the newest multilevel sub-20-nm technologies while read/write latency is considerably higher than a DRAM. This means that NAND is not used as a working memory, but, rather, the architecture that became the mainstream of electronic systems is made of a bank of NANDs that permanently stores code and data that are not frequently changed, e.g., some parameters used for system settings, while a DRAM is used as a working memory. An example of this architecture for a consumer product is shown in Fig. 1.

The growth of applications and in parallel of the data to be stored and processed forced the continuous improvement of NAND technology to increase the storage capacity. This means, essentially, to reduce the dimensions of a memory cell through a technology scaling roadmap that brought NAND down to the 16-nm technology node as shown in Fig. 2. In addition, years ago MLC was introduced at that time to increase the density in the same technology step.

In doing so, however, endurance has deteriorated, and the signal margin for read has been further reduced, increasing the importance of some fundamental limits that have the power to slow down and eventually stop the NAND scaling path.

The first issue is the reduction of stored charge due to the reduction of cell capacitance: for state-of-art—technology, stored charge is on the order of tens of electrons (see Fig. 3), which causes a large fluctuation in the threshold voltage shift representing the detectable signal in the memory cell. In addition, the shrinking of

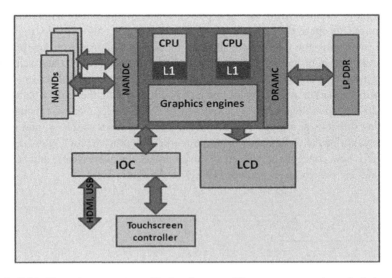

Fig. 1 Tablet/Smartphone system block diagram. The memory system includes fast LPDRAM DDR as working memory and a bank of NAND's for mass storage

Fig. 2 Evolution of NAND cell size

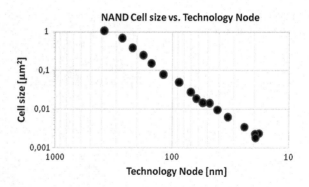

Fig. 3 Down the scaling path, V_t shift per electron reduces drastically

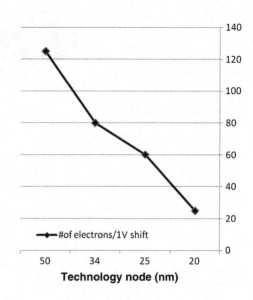

dimensions produces an increase in cell-to-cell interference with a stronger impact for sub-30 nm technologies. This effect is caused by the fringing capacitance between one cell and the next one in the same row, which results in a shift of threshold voltage (Fig. 4). The combined effect of these two phenomena produces a widening of threshold voltage (V_t) distribution that makes it more difficult to correctly read the memory.

The second problem is that a wide spacing between V_t distributions (V_t window) is hard to maintain with scaling (Fig. 5).

Consequently, the scaling path produces the result of a narrower V_t window and, at the same time, a widening of V_t distribution that increases the error rate during reading. Ultimately the V_t margin disappears with scaling (see Fig. 6).

Scaling roadmap requires also the reduction of cell's gate oxide, but gate oxides are facing increasing limits to further reduction in thickness for two reasons: the first one is the onset of direct tunneling for $t_{ox} < 5\text{–}6$ nm. This means that current

Fig. 4 Modeling of
cell-to-cell interference

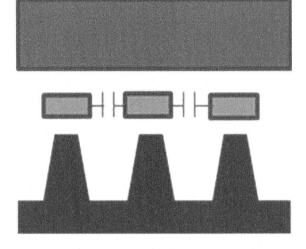

Fig. 5 Widening of V_t
distribution with reducing cell
feature size

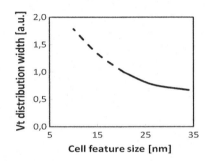

Fig. 6 V_t window narrowing
with cell scaling

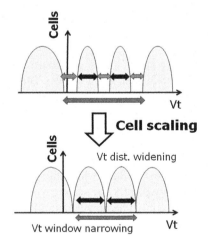

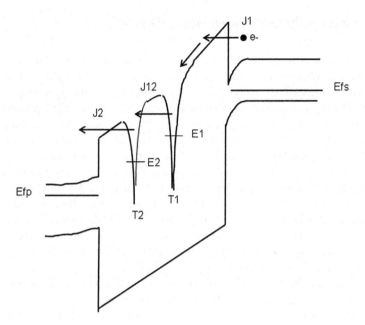

Fig. 7 Trap-assisted conduction

can flow through the oxide due to the tunnel effect, but without help from an electric field, worsening the retention characteristics of the cell. The other point is the onset of a trap-assisted leakage current for $t_{ox} < 8$–9 nm, as shown in Fig. 7, in particular after repeated write/erase cycles. Both these effects result in a worsening of retention characteristics and endurance. In the end, the limit to reduction in gate oxide thickness results in high voltages still to be used during program and erase. The final stages of decoding circuits cannot be scaled because they have to accept such high voltages, and this is a constraint to the size scaling of the array.

There are also other issues connected to some technology steps: Isolation is more challenging because distances are scaled down, but voltages aren't, so shallow trench isolation (STI) tuning is needed to prevent parasitic leakage paths, and also there are oxide filling issues due to high aspect-ratio morphology.

The floating gate is another critical aspect: the coupling ratio is degraded strongly impacting cell performance.

The contact to drain select transistors is more challenging as the reduced pitch results in a very high aspect ratio. Finally, metal properties impact bit-line resistance and capacitance.

In the attempt to overcome the increasing scaling difficulties 3D architecture was recently introduced for NAND and used with MLC.

3   Exploiting the 3rd Dimension: 3D NAND

The 3D concept is to use the third dimension in the silicon to allocate memory arrays via a number of stacked layers, so that the memory density is multiplied by the number of stackable layers without the need to scale the technology. A 3D NAND can offer a way to continue along the cost-reduction roadmap without the scaling cell size magnifying the problems briefly discussed above. In Fig. 8, the effective cell size is compared with the physical cell size during the transition to 3D NAND, showing a relaxed feature size of more than a factor of two.

The effect of this relaxed technology on the main issues encountered in NAND-cell scaling path is that the number of electrons per 1 V of V_t shift comes back to the level of a much relaxed technology, while the floating-gate to floating-gate interference is lower. In the end, 3D NAND provides narrower V_t distribution width that is able to operate safely even with multilevel architecture. The distribution width can be about ¼ of the value obtained with 2D NAND at the same effective cell dimensions.

3D-NAND architecture can be divided in two classes: vertical 3D and horizontal 3D. Figure 9 depicts a schematization of the two options.

In vertical 3D, the channel of NAND cells is made of dots that cross the layers of the gates, and the channel length of each transistor of the NAND string is determined by the thickness of the poly layer, while channel width is determined by the perimeter of the dot. To minimize the occupied area, the diameter of the dot is defined by the process-feature size. Gate layers are common to all the cells of a bit-line and are not lithographically defined. On the contrary, in the horizontal approach the gates rise vertically and are defined to build gate length, while channel runs horizontally. In this case, channel width is defined by the thickness of the channel layer. Examples of both approaches have been developed and will be described in the following paragraphs. In 2007 BiCS approach was introduced [2]. This is a vertical 3D architecture with a SiN trap cell, which can be schematically depicted as in Fig. 10.

Fig. 8 Advantage of 3D in achieving compact cell size with relaxed technology feature

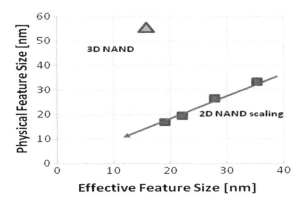

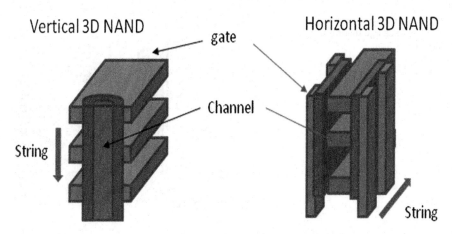

Fig. 9 Vertical and horizontal are complementary approaches to a 3D array

Fig. 10 BiCS architecture

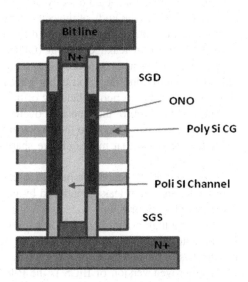

A polysilicon channel is contacted at the bottom with an N+ source line and at the top with a metal bit-line. SGD and SGS are the two selection transistors at drain and source level respectively, while the stacked polysilicon layers form the control gates of the cells in the NAND string. ONO is the gate insulator where the stored charge is trapped. This structure doesn't require definition of the polysilicon to obtain the right channel length for transistors in the NAND cell, but rather it requires opening a hole with very high aspect ratio. ONO layer cannot be cut from one cell to the other in the string. High compactness is ensured while a potential drawback is that some charge migration between the neighboring cells can take place, leading to retention reduction and charge trapping between the cells that may degrade cycling performance.

Fig. 11 TCAT solves the
problem of charge sharing and
trapping

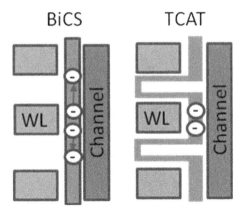

A 3D approach challenge is evident during erase operations: the 2D-NAND cell's body can be connected directly to high voltages (~ 20 V); on the contrary in 3D-NAND devices the body of the cell is floating and the bias voltage for erasing is supplied through the common source diffusion using the Source Gate Selector to control the junction leakage current (GIDL effect). In fact GIDL current of the source selector transistor supplies holes required for erase, but this phenomenon is difficult to control and may lead to hot carriers damage, high cell-to-cell variability and slow erase due to poor efficiency of this mechanism.

Another architecture that can also overcome the issues mentioned above is the TCAT approach [3]. Thanks to the "replacement gate process", the SiN layer has a characteristic shape that makes charge sharing and trapping between cells impossible (see Fig. 11).

Also, the pillar is directly connected to the substrate so that a conventional erase is possible, and drain leakage induced by gate bias (GIDL) is no more required to generate holes for erasing.

There is a high price to pay for these improvements: it is the extra space needed for the gate process and the body contact. It has been calculated that a 30–40% penalty has to be expected.

An improved version of BiCS is the P-BiCS [4] introduced in 2009. This is a vertical U-shape pillar in which the source is contacted via a metal line on the top that solves some of the problems of the simple BiCS. In particular, write doesn't need GIDL anymore. Even in this case, a certain area-occupation penalty (about +20% on BiCS) must be paid for this approach.

All the approaches treated so far are in the category of vertical 3D in which the channel is running vertical in the array and control gates are on stacked layers crossing the channel. Another option, in which the channel is running horizontally and the gate lines are running vertically, has been explored in 2009 and 2010 with the so-called vertical gate NAND (VG-NAND) [5, 6].

VG-NAND achieves a smaller cell size with respect to the vertical option because of the lack of gates all around the channel, but lacks the shielding effect

Table 1 Summary of strengths and weaknesses of most important 3D-NAND architectures

	VG-NAND	BiCs	P-BiCs	T-CAT
	Horizontal	Vertical	Vertical	Vertical
Pros	Footprint (++)	Footprint (+)	Low gate resistance	Low gate resistance
			P-bulk erase	Charge sharing
Cons	Cell-to-cell interference	Charge sharing	Charge sharing	Footprint (− −)
	Poor coupling ratio	GIDL erase	Footprint (−)	

offered by surrounding gates, so this may lead to higher cell-to-cell interference. Also, the gate-coupling ratio is worse than in the case of vertical 3D, and W/E performance may be affected. To try to summarize what has been said in this quick overview, Table 1 lists the pros and cons for 3D NAND architecture is presented.

4 Breakthrough Approach: Emerging Memories

In the previous section, we have seen a short summary of the powerful efforts made to avoid the slow-down of size reduction rate of NAND memories imposed by technology limits [7]. At the same time, the availability of new materials whose resistance can be programmed to two (or more) well separate values opens the possibility of a replacement for NAND by a completely different technology, but a big push in this direction came from a deep rethinking of the old, established system memory based on DRAM, NAND, and disks that started because it had been realized that maintaining the performance growth rate required by state-of-art computer systems and by the highly demanding applications was becoming a major challenge [8]. On one hand, the gap between computing and memory speed which is already five orders of magnitudes is rapidly increasing together with energy consumption, space requirements, and cost of memory. On the other hand, critical applications are yet undergoing a paradigm shift, from the old computer-centric paradigm, based on CPU performance, driven by computational applications like computer-aided design, physical problem solving, and mathematical calculations, to a new data-centric, paradigm driven by social media, e-commerce, e-finance, search and mining, and digital media communications which require processing petabytes of data. An effect of multiplication, due to the fact that beyond each byte of useful data it is necessary to transmit additional bytes to make it possible, data processing tends to expand even more this already incredible quantity of bytes needing storage. It has been calculated that this multiplication effect can be also in the range 1000:1.

Figure 12 schematizes the evolution of memory-hierarchy architecture. The top row sketches today's state of the art of the gap in terms of latency between the mass storage devices (tape and disk) and the rest of the memory system nearer to CPU (cache and DRAM), which is on the order of 10^6. Mass storage is characterized by non-volatility and big memory size, but the R/W bandwidth is not enough to be

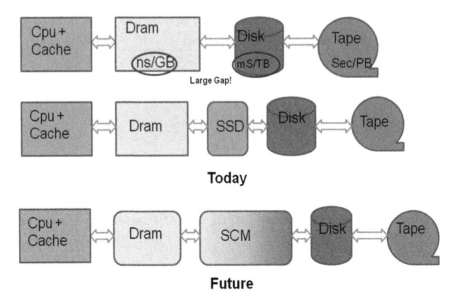

Fig. 12 Evolution of memory-hierarchy architecture

used as a working memory. So the data cannot be processed directly on the non-volatile storage devices, but part of them have to be loaded into fast volatile memory to be processed and to be then stored again in the non-volatile section of memory hierarchy, thus generating many R/W transfers, which slows all the processing system: this is a bottleneck that state-of-the-art systems cannot afford anymore. As we have seen in the above discussion that a big amount of data has to be processed in new applications, it needs to be processed in real-time: in e-trading, for example, a minimal delay in response can cause a million-dollar loss, while, on a completely different side, complex systems like electric-power-network controls, life-saving medical equipment, and on-board aircraft systems have to recover their status immediately after an accidental shut-down to avoid catastrophic consequences.

A step forward to mitigate this problem has been made possible by the availability of large-sized NAND memories, both with MLC and 3D architecture. With these devices, it is possible to build solid-state disks (SSD) at GB scale, with latency on the order of μSec, to be interposed between DRAM and HDD. In spite of the fact that this solution can reduce the size of HDD or even eliminate it and speed-up the computer boot and operation, latency remains too high to enable the direct execution of stored code, so data processing is still performed in DRAM, and the frequent downloading from and to SSD is a significant challenge for NAND endurance.

To overcome these obstacles and maintain the past growth rate in large-scale memory systems, we can look at the opportunities opened by emerging memory technologies to create a new type of memory, the so-called storage class memory (SCM) [9, 10] able to reduce the performance gap between fast, expensive, and volatile DRAM and slow, cheap, and non-volatile mass memory like NAND in SSD and HDD. SCM has the potential to expand up to DRAM size and, in the opposite direction, the SSD/NAND side, reducing their installed capacity.

The defining requirements for all the technologies intended to be used for an SCM are:

- High R/W throughput: on the order of 1 GB/s.
- High write endurance: much better than that delivered by NAND, on the order of 10^{12}.
- Low read latency: on the order of 100 ns.
- Non-volatility: no or only sporadic need for refresh of stored data, i.e., ideally the 10-year retention that flash technology can offer.
- A good scaling perspective, with the potential to overcome limits that are currently restricting NAND and DRAM scaling perspectives.

In addition, an SCM offers mechanical robustness (no moving parts) and much lower energy consumption: non-volatility with low latency may encompass the possibility that a DRAM-like device will not consume the power now spent performing periodical refreshes.

The early idea to develop a technology for a "universal" memory capable of meeting all the requirements today supplied by SRAM, DRAM and FLASH has proven to be not easy to realize.

It is interesting to note that not necessarily the one same technology must meet all these requirements, but this rather may depend on the development targets.

In recent years, industrial researchers have identified various types of appealing technologies, some with characteristics more suitable for NAND replacement, others more similar for DRAM.

Among them, the most mature is Phase Change Memory (PCM) whose development started around 2000 and is now at a production stage, but also "old" technologies, widely used in small market niches such as MRAM and FeRAM that found unexpected opportunities due to advancements in concepts and new materials. Finally, a large family of memories based on conductive oxides, namely ReRAM, gained attention due to its relatively simple technology and high scalability.

Some of these technologies are suitable for a 3D integration, which reduces the gap with the major competitors, but NAND can also be Multilevel, and the combination of these capabilities is a big challenge for the emerging technologies proposed so far.

4.1   Phase Change Memories (PCM)

PCM are based on the change of resistivity from low to high of chalcogenide alloys depending on their crystalline or amorphous status [11]. Chalcogenide became very popular as a storage layer on CDs and DVDs: in this application, the memorization of the information was done using a focused laser beam. The energy of the laser could locally melt the material causing a transition from crystalline to amorphous and a change of material reflectivity that could easily be detected [12].

Later, chalcogenide were used to build a non-volatile, solid-state memory [13] capable of competing and even exceeding the performance of mainstream flash devices. Writing to this new memory was faster than a flash, and even a single bit could be modified independently, so that there was no more need to erase a full block of cells in the array. An electrical current was used to change the phase of the material, thus recording data into a memory cell, and this data remains stored even if the power supply is removed from the device, until another current pulse is able to write new data.

PCM is the most mature among the new memory technologies, and, as of today, memory devices at the Gb level in 40-nm technology are in production [14].

Figure 13 shows a simplified cross section of a PCM memory device: a layer of GST material (see chapter "Physics and Technology of Emerging Non-Volatile Memories") shown in its crystalline phase is deposited above a pillar of resistive material, and the entire structure is sandwiched between two electrodes. The role of the resistor (heater) is very important as it generates the heat to reach a high temperature in a small area in which resistor and GST layer are in contact, called active region in the figure. Design of the heater is very important for the power efficiency of the cell because there are some key points that must be fulfilled to obtain the best results: temperature peak must be near the interface between GST; and heat diffusion from heater's lateral surface must be minimized, not only due to efficiency reasons, but also because of the possibility of disturbing neighboring cells in an array and spuriously causing reprogramming. Active regions can be made crystalline or amorphous, applying a suitable current pulse across the material. In

Fig. 13 Schematic-process
cross section of PCM cell

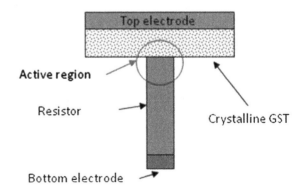

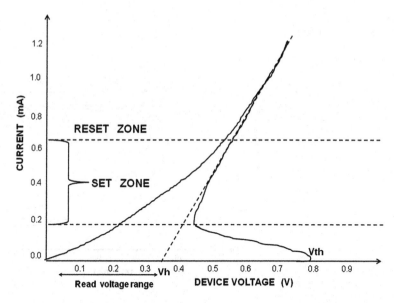

Fig. 14 Static I/V characteristic of a PCM cell. The switch back of the characteristic of amorphous bit at V_{th}, when transition to crystalline state takes place, should be noted

the first case, the cell is called SET and shows low resistivity; in the second case, it is called RESET and shows high electrical resistivity. The difference between these two conditions is three or four orders of magnitudes, making it possible to use the material as a memory device having a large signal available to read the stored data, which is a critical requirement to obtain high read speed and low error rate. Eventually, intermediate levels could be achieved in the window between low and high resistivity, making it possible to store more than one bit in a single physical cell (MLC).

In Fig. 14 is clearly visible the hysteresis due to the two possible states in which the memory cell can stay and we will explain it to understand the cell's operation. Starting with an amorphous material and moving up the axis of applied voltage, the current remain very low until a critical voltage is reached (the threshold voltage V_{th}). At this point, an electronic conduction phenomena is triggered by the high electric field applied to the chalcogenide material, which has the effect of lowering the threshold to conduction (threshold switching) and the material switches to a state of high-conductivity threshold. In the first phase of the switching process, the conduction is dominated by electronic conduction, and the process is reversible: if the voltage is removed at this point, the material will return to a high-resistance state [15]. On the contrary, in the second phase the high current flowing into the heater produces, due to the Joule effect, a large temperature increase that leads to structural, permanent modifications of the material. After the snap-back, the I/V characteristic is located on the dashed line visible in Fig. 14. Ideally this should be a perfectly vertical line crossing the x-axis at a point called V_h (V_{hold}), which is a

Fig. 15 Degree of transformation of GST from amorphous to crystalline as a function of time and temperature

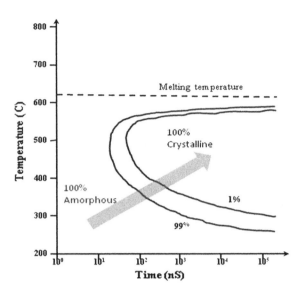

characteristic holding voltage depending on the material, but, in the real structure, the contacts to the chalcogenide material and other characteristics due to the structure itself introduce a resistance that is defined as the finite slope of the dashed curve in Fig. 14. To reverse the process and initiate the structural modification of the material to make the transition from amorphous to crystalline phase, a critical temperature called the crystallization temperature (T_{cryst}) must be reached. This temperature depends on the material and it is a critical parameter because: if the temperature is too high the material remains amorphous, if it is too low, nucleation of microscopic crystals that are the starting point for the macroscopic crystallization process is extremely slow, as shown in the drawing of Fig. 15.

The curves drawn are for 99 and 1% degrees of remaining amorphous material. From the figure, the existence of an optimum crystallization temperature and then an optimum current pulse, at which the crystallization speed is higher, is clear.

It should be noticed that the amorphous phase is not stable and tends to evolve into crystalline phase which minimizes the free energy of the system, but high activation energy allows amorphous phase to stay practically unchanged for years at room temperature. This is one of the most important features of PCM, which enables it use as for non-volatile memory.

The portion of the I/V curve below about 0.5 V is the region used to read the PCM cell. Usually this is done, at least in principle, by detecting the variation of a bias voltage applied to the cell, modeled as a nonlinear resistor, caused by the application of a known current. The right voltage to be applied is a compromise between the possibility of having a larger signal and a faster reading and the risk of disturbing the cell by spurious programming; in general, voltage bias during read is <0.4 V.

Fig. 16 Pulse shape for SET and RESET

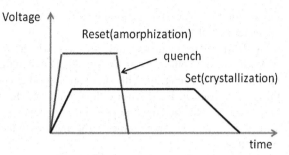

Starting from a crystalline material, the amorphous phase is reached by applying a current pulse high enough to melt the material, to recover crystalline state it is necessary a pulse with the same polarity, but lower value, because a lower temperature is needed for crystallization.

Let's now have a closer look at the pulses applied to GST material in the various conditions and let's name the pulse setting a low resistance phase in the material as SET, while the one restoring the high-resistance phase as RESET. With reference to Fig. 16, where the horizontal axis represents the time and the vertical axis the height of the pulse in volts, the RESET pulse must be capable of raising the temperature above melting point and its width is on the order of 20–50 ns. But the key point is that the trailing slope must be very sharp (about 2 ns), because once the material has reached the melting condition it has to be quenched to "freeze" the amorphous phase.

This is a consequence of the fact that amorphous is not the stable state for GST, and, if the temperature would be reduced gradually, the re-crystallization process could begin. On the contrary, the SET pulse height is lower because we have seen that it must reach a lower temperature and its duration is considerably longer (150–300 ns) than RESET and this is so because a longer period at lower temperature and a smooth decrease is required to enable nucleation and growth of the crystalline material.

Finally, the read operation is helped by the large resistance difference (wide signal), and latency is on the order of 30–50 ns, comparable with DRAM and flash NOR.

Due to the program concept involving the Joule effect, the power per bit required by PCM to program is not negligible, e.g., regarding the reset case, which is the most expensive, it is given by:

$$P_{reset} = V_{hold}\, I_{reset} + R I_{reset}^2$$

Where R represents the slope of the I/V curve caused by contact resistance and the resistance associated with the heater, the energy for the programming operation is on the order of 100 pJ/bit and, due to this, it is not possible to write many cells in parallel. This is a limit in the write throughput which make it more difficult for PCM to compete with NAND, and the SET operation is the limiting factor because of the longer duration. The SET duration is affected by the crystallization speed which is a characteristic of the material.

A PCM cell is made not only of GST which composes the variable resistor. A device must be put in series, i.e., ideally a switch that enables the selection of the particular cell in the array, without disturbing other cells. Actual implementation of this ideal switch in PCM can be done with an MOS, a BJT, or a diode or other non-linear I/V characteristic device, but, whatever is the choice, this device has a critical role in determining the performance of the PCM cell and the scaling perspective.

The scaling perspective of PCM is also connected to the capability of the selector device to provide the current needed for programming, mainly RESET which requires the higher current and in parallel depends on the ability to reduce the SET/RESET currents with scaling. When actually considering an isotropic scaling of a factor K (K > 1), the behavior of main parameters is:

Current density: $J \sim K$
Heater contact area: $A \sim 1/K^2$ then,
Current: $I \sim 1/K$
Voltage: $V_{cell} = const$
Power dissipation: $Pdiss \sim 1/K$

The good news is that the writing current scales with the area and the same happens for power dissipation, this is one of the most important features of PCM.

We can now have a look to the complete memory structure: cross sections of a MOS and vertical bipolar cells are shown in Figs. 17 and 18 with their circuital schematics shown in Fig. 19.

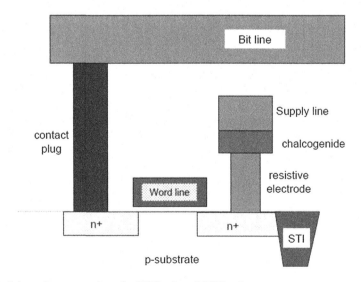

Fig. 17 Schematic cross section of a MOS selected PCM cell

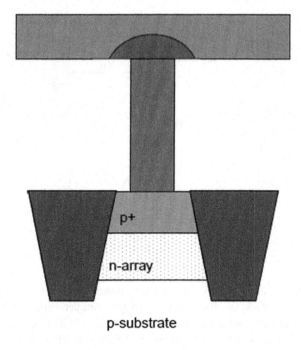

Fig. 18 Schematic cross section of BJT selected PCM cell

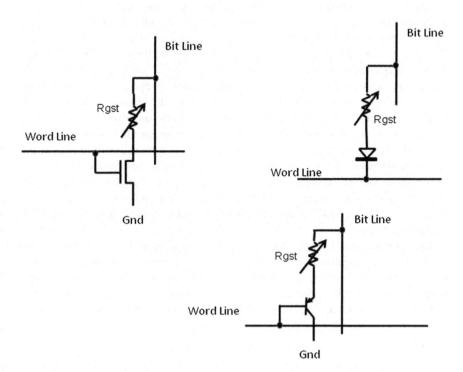

Fig. 19 Equivalent circuits of a PCM cell with different selectors

Fig. 20 Distance Xd is enough to reduce the temperature to non-critical values even at very small technologies (20 nm)

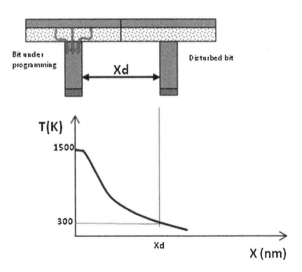

MOS selector allows array architecture more similar to the well-known flash-NOR, thus making the circuit design simpler and optimizing the array-leakage currents, but a larger cell is expected (15–20 F^2). The best performances for cell area (4–6 F^2) and current capability are obtained with the vertical diode. Bipolar transistor (BJT) also has been adopted as a selector in some PCM designs.

Another issue depending on scaling capability is the thermal cross talk, because the heat necessary to write a cell cannot be completely confined to the selected cell, but can spread through the chip to affect neighbor cells (see Fig. 20). For example disturbed amorphous cell can suffer accelerated crystallization with consequent drop of electrical resistance and loss of stored data. This risk is greater with technology scaling and with an increasing number of cycles, but experimental results show that thermal cross-talk does not affect cells up to 20-nm technology until it has reached a very high number of cycles. (on the order of 10^{10} cycles) [16].

We have previously explained that the stable phase for GST is the crystalline one, and, even if we "freeze" the material in the amorphous condition, under the effects of temperature and time, the material will tend to be restored to the crystalline phase over an extended period. Of course this is an issue because this results in a data retention problem for PCM. The crystallization tendency is a phenomenon highly sensitive to temperature, and retention decreases very quickly with temperature increases.

Retention can then be a problem for highly demanding temperature applications, or for those applications in which the memory content is written before the assembly process. High temperatures used during the assembly flow can in-fact affect memory content. Improvements to overcome this issue are on the way; from a technology viewpoint, different GST alloys that exhibit better retention

Fig. 21 Cross section of a
generic Re-RAM memory cell

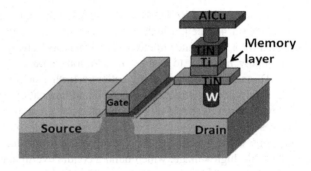

characteristics with temperature need higher currents to RESET, so higher power
consumption, while, on the other hand, design solutions such as sporadic refresh
could be considered.

4.2 Resistive RAMs (ReRAM)

As mentioned in chapter "Physics and Technology of Emerging Non-Volatile
Memories", the term ReRAM indicates a quite large family of devices based on various physical concepts [17–19]. In all the cases, the memory cell is made of a sandwich
of the memory material between two electrodes of conductive material, in series with a
selector device as in Fig. 21; the obviously planar MOS selector shown in this figure is
not suitable for high-density memory and must be replaced by another selector device.

Instead there are remarkable differences between the physical principles and the
biasing during programming: the first classification from a circuital viewpoint can
be done distinguishing between unipolar and bipolar devices. Unipolar devices can
be programmed in both their states by changing the level of applied voltage but not
its polarity. Bipolar devices instead need to reverse the polarity of applied voltage to
change the state of the memory material, and, in spite of the more difficult decoding
at the circuit level, they have been preferred because of the lower voltage needed for
SET/RESET, lower power consumption, and higher speed.

A further classification among bipolar ReRAMs can be made based on the
conduction mechanism, i.e., filamentary or interfacial.

In filamentary devices current conduction takes place through a
microscopic-level path opened in a dielectric, they offer large scaling potential
because the filament dimension is much smaller than the memory stack area, and it
is not affected by dimension reduction. On the other hand, optimization of the cell's
current is more difficult, and the behavior of a cell tends to change from one cycle to
another. They also have better characteristics of retention and much better endurance, while interfacial devices, based on uniform movement of O^{2-} ions have been
shown poor retention. Interfacial devices have shown limited endurance and limited
scaling capability in addition to a low read current [17].

A strong point of Re-RAM technologies is that most of them have the potential to achieve a $4F^2$ cell, which positions them very well for scaling. They can be stacked for a 3D architecture to enhance memory density, and MLC is feasible even if probably not easy to realize. In addition, endurance can be much better than NAND, and bit-level granularity offers a further advantage for software architecture, in addition the structure of the cell and fabrication process is relatively simple.

Oxide based Re-RAM [20–22] has attracted great interest from industry. In these devices, conductivity into an oxide is sustained by oxygen vacancies (oxide bond locations where the oxygen has been removed) created after the application of a sufficiently high voltage, which build a conductive filament which can be used as a bridge by electric charge to flow into the oxide. Generally a metal oxide like TaOx, TiOx, ZrOx or HfOx can be used as a dielectric layer. The basic structure of a typical cell is MIM (Metal– + Insulator–Metal) as shown in Fig. 22 where top electrode is TiN/Ti and bottom electrode is Ti with a very thin oxide, (HfOx)on the order of 10 nm, but other choices are possible; for a simplified picture of the conduction mechanism, see Fig. 23. During the set process, the oxygen atoms are pulled from the lattice under the effect of a large enough electric field, and, as the percolation paths of oxygen vacancies are formed between the two electrodes, the oxygen negative ions move in direction of positive voltage and get stored at the interface between electrode and oxide, before the device goes into the low-resistance state. On the contrary, during the reset process, oxygen ions are driven back and recombine with the oxygen vacancies under the effect of a reversed electric field, and, as a result, conductive paths are partially ruptured as the device goes back to the high-resistance state. This process is illustrated by the I/V characteristic of Fig. 24: operation of the cell is in principle simple, starting from the lower curve (high resistance) and applying enough voltage between top and bottom electrodes, we can switch to the top curve (low resistance), called the SET as in PCM. On the other hand, by applying enough voltage of opposite polarity, it is possible to switch back to the starting curve, and this is called RESET. Generally current compliance is provided during a SET operation to avoid stress on the cell

Fig. 22 Cross section of an oxide-based ReRAM cell

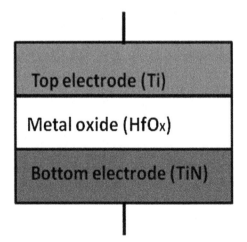

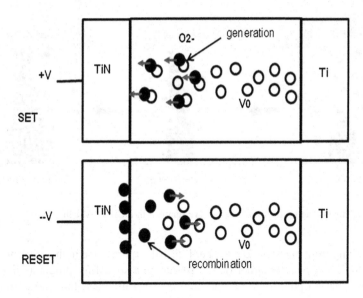

Fig. 23 Set/reset mechanism

Fig. 24 Schematic I/V
characteristic

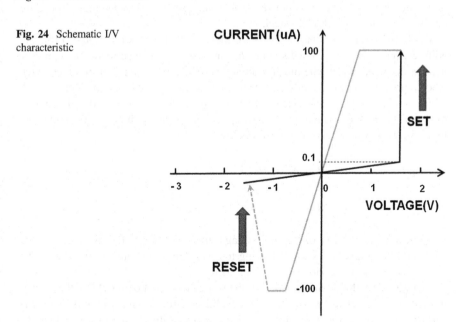

caused by a sudden high current flow, and a voltage limit is provided for RESET
[23, 24].

An appealing feature of this device is the low voltage and current required to
perform SET/RESET.

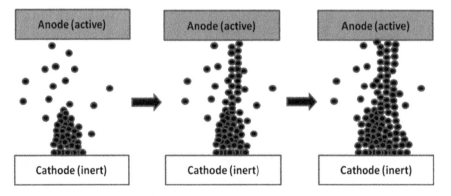

Fig. 25 Formation and growth of a conductive filament in a CB-RAM cell

Switching from one state to the other [25, 26] requires time in the order of 10 ns, so roughly energy on the order of 1–5 pJ can be assumed for programming.

From a design point of view, the ratio between high and low current must be considered, as it defines the magnitude of the read signal, this parameter is reported on the order of 10. Retention and endurance can vary over a wide range depending on the material choice for the stack.

CB-RAM is another class of resistive memory that works on another principle with respect to those examined so far. They are based on an electrochemical process that oxidizes an electrochemically active metallic anode (e.g. silver or copper), when a positive voltage is applied. M+ cations drift within an ion-conducting layer which is a thin film of solid electrolyte, reducing at the relatively inert cathode (e.g. tungsten or nickel) which leads to deposition of a conducting filament (See Fig. 25). This process is schematized below:

$$M^0 - e^- = M^+ ;$$
$$M^+ \rightarrow \text{cathode}$$
$$M^+ + e^- = M^0 \quad \text{(Filament growth)}$$

Inverting the polarity of applied voltage, the process can be inverted, and the conducting filament is destroyed leading to a very high resistance of the stack. The I/V characteristic is schematized in Fig. 26.

During set, with the anode acting as the source of metallic ions (M+), they travel through the dielectric layer, which is a solid electrolyte and accumulate on the surface of cathode for a reduction and electro-crystallization process; when the filament is formed (ON state), it grows under the effect of electric field.

Reversing the applied voltage, anodic ionization and M+-ion reverse transport partially dissolves the filament (reset). Eventually, the device turns completely off. During the initial phase when the filament is still connected, the ionic and electronic currents are present together, generating a large reset current.

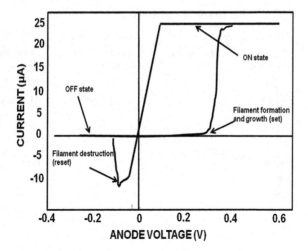

Fig. 26 Schematized I-V characteristic of a CB-RAM cell

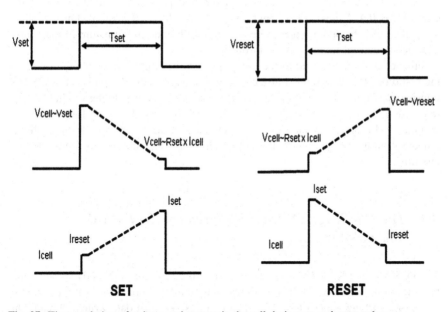

Fig. 27 Time-evolution of voltage and current in the cell during set and reset pulses

Another important point for designers is to establish current compliance during the SET operation to be able to set low resistance Rlow at a correct value.

The evolution of voltage and current through the cell during SET and RESET is schematized in Fig. 27. The pulse width needed to SET and RESET is larger than that in oxide- based memories, and the energy involved is higher, in the range of 100–200 pJ.

Table 2 Summary of Re-RAM family characteristics

	Filamentary (bipolar)	Interfacial (bipolar)	CBRAM (bipolar)
Operation principle	Oxygen vacancy filament in dielectric	Uniform oxygen movement	Metal filament in electrolyte
I_{on}/I_{off} Ratio	10	10	1000
Operating current	25 uA	Area dependent	<100 uA
Speed	<10 nS	1 uS	100 nS
Endurance	1e10	1e4	1e7
Retention	Good	Fair	poor
Scaling/3D	Fair	Fair	Fair
MLC capability	Fair	Good	Good

We can better appreciate the differences between the main characteristics of these approaches to Re-RAM as summarized in Table 2.

Filamentary bipolar devices seem to have better speed and endurance, and, in the case of CB-RAM, also a very high signal available (I_{on}/I_{off} ratio). On the other hand, the cell's behavior tends to be not repeatable, due to the nature of filaments, and the set current is not dependent on the scaling.

Filamentary devices tend to require a higher voltage pulse to create the filament the first time. This "forming" pulse creates some issue in the design because CMOS transistors driving the array must support a considerably higher voltage than required over the rest of device's life, in addition the choice of forming pulse voltage level is driven by the worst-case cell and that means an over-stress on all the other cells in the array. Finally, this operation requires an addition to overall device test time.

4.3 The Challenge of Dram Replacement: STT-RAM and FeRAM

The PCRAM and Re-RAM approaches can be used to replace NAND storage, due to their ability to scale and the possibility of a multilevel operation, but they cannot cover the working area of DRAM which requires very fast data write and read. On the other hand, the dream of universal memory is to replace in the memory hierarchy both NAND and DRAM, at least for most applications [27, 28]. The emerging memory approaches we will describe in this chapter have much better chances to span from DRAM- to NAND-like performance, and they could result in the best compromise for a real universal memory.

The history of magnetic RAM is relatively long. The first non-volatile memories used in computers were based on modification of the magnetization of small ferrite

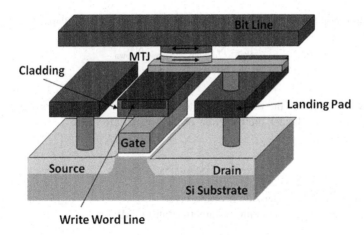

Fig. 28 Example of a structure for early magnetic memory

Fig. 29 Schematic of an MTJ

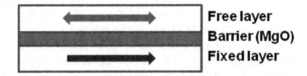

devices using a strong electric current flowing through them. In recent times, the same concept was used to design a non-volatile semiconductor memory whose structure is shown in Fig. 28.

The heart of a memory cell is a structure called a Magnetic Tunnel Junction (MTJ) [29, 30]. It can be schematically described as a thin insulation layer sandwiched between two ferromagnetic layers, one of which, called the pinned or fixed layer, has been strongly magnetized in one direction, while the other, called the free layer, has weaker magnetization whose direction can be changed under defined conditions (see Fig. 29).

In Fig. 29, the red line with double arrows is used to indicate that the direction of magnetic field in the free layer can take either a direction parallel (P) or anti-parallel (AP) with respect to pinned layer. The insulator barrier is very thin (on the order of 1 nm), so that a tunneling current can flow in the structure when an electric field is applied to the electrodes. The interesting property of this structure is that, when the direction of magnetization of the two layers is parallel, much more current flows through the structure. This behavior is called the Giant Magneto Resistance effect (GMR), and it is dependent on the energy distribution of electronic states on both the ferromagnetic electrodes which is affected by the interaction with the magnetization direction.

The tunneling current can be expressed with the following relationships

$$I_P = N_{up}^1 N_{up}^2 + N_{down}^1 N_{down}^2$$
$$I_{AP} = N_{up}^1 N_{down}^2 + N_{down}^1 N_{up}^2 \tag{1}$$

Where N indicates the density of states at Fermi energy in layers 1 and 2 and "up" or "down" indicates the direction of the spin of electrons.

The quantity:

$$P = N_{up} - N_{down}/N_{up} + N_{down} \tag{2}$$

Indicates the spin polarization of the layer.

We can define the Tunneling Magnetic Ratio (TMR) as:

$$TMR = (Rap - Rp)/Rp \tag{3}$$

where Rap = anti-parallel resistance and Rp = parallel resistance.

This number is a measure of the difference of resistance between the two possible states of the MTJ relative to parallel resistance. Normally, it is given as a percentage, and it is a measure of the available reading window of the memory cell, as shown in Fig. 30. This is also the reason why a great effort is being put by technologists to increasing this parameter as much as possible. Today's (2015) state-of-the-art values are in the range 100–150%.

In the first approach to MRAMs (Fig. 29), the cell is organized with a selector transistor connected to one of the electrodes, while the MTJ is placed at the crossing point of the bit-line wire and an orthogonal line called the write word-line. Another word-line is connected to the MOS device to select the cell. To change the

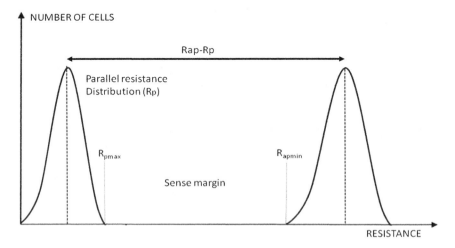

Fig. 30 Tunneling Magnetic Ratio defines the available reading window for MTJ

magnetization of the free layer, it is necessary to select the appropriate bit-line and the write word-line. The sum of the magnetic fields generated by these two "wires" is enough to orient the magnetization of the free layer in the desired direction. Selectivity of the modification is guaranteed by the fact that only the cell in the position to be affected by both the magnetic field generated by bit-line and write word-line can be modified.

While the first MRAMs exhibited bit-level granularity fast and symmetrical write/read, very good endurance, and low-voltage operation, they had a number of drawbacks:

- Cell area is large due to double word-line and the need to have the MTJ position aligned with the crossing of write word-line and bit-line, plus the difficulty to be scaled down.
- High current is needed to create an external magnetic field able to reverse free layer magnetization. This leads to significant metallic and electromigration problems.
- Even if only the sum of two perpendicular fields is able to switch magnetization of the free layer, the field cannot be perfectly confined and disturbs adjacent cells during write which can become an important issue.

This MRAM technology has been used for embedded applications and limited memory capacities.

The discovery in 1996 [31–33] of the Spin Transfer Torque (STT) effect changed completely the perspective, and magnetic memories became appealing also for high density memories [34].

The spin-Transfer-torque (STT) principle makes use of magnetic tunneling junctions, so the insulating barrier between the magnetic materials can be a metallic oxide. Early MRAMs used Al_2O_3 but today MgO is used because higher TMR and a lower write current can be obtained.

Compared to the first MRAMs, STT-MRAM uses the same current flowing through the tunneling junction to flip the magnetization of the free layer and consequently switch the device resistance from high to low and vice versa. To achieve this result, STT is used; an exact description of this effect is beyond the scope of this book, but we provide here an approximate description of how it works with reference to Fig. 31. The magnetization M_1 of the fixed layer acts as a filter for the spin of electrons flowing through it, so that, at the output, we have a current of electrons with spin oriented in the direction of M_1 (Fig. 31a). When they interact with magnetization M_2 of the free layer, they start to precede around the M_2 direction and this means that spin transversal component (S_y) matches the direction of magnetization M_2.

But, due to the different speed and direction, each electron spin will precede with a different angular speed and, after a short distance; the electrons will be completely uncorrelated with the resulting transversal spin component (S_x) of the population equal to zero (Fig. 31b). On the other hand, the spin magnetization must be preserved so that the transversal component is transferred to the magnetization of the

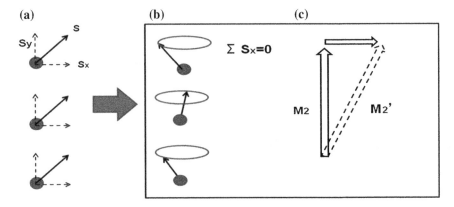

Fig. 31 Schematic of STT (Spin Transfer Torque) mechanism

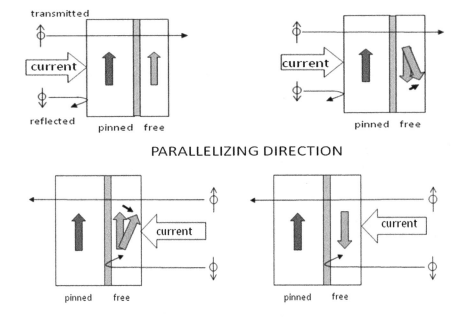

Fig. 32 In an STT device, one current direction tends to parallelize M_1 and M_2, while the other tends to anti-parallelize them

free layer as a torque, and, under the action of this torque, if it exceeds a critical value, magnetization M_2 can be reversed (Fig. 31c). Actually the result of the interaction between electrons and magnetization of the free layer depends on the direction of the current. One direction tends to parallelize M_1 and M_2, while the other tends to anti-parallelize (see Fig. 32).

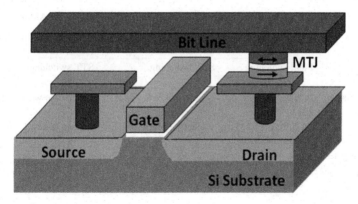

Fig. 33 Cross section of a STT-RAM cell

In the upper part of this figure, electrons are flowing from the left, and the pinned layer works as a transmission filter, as electrons with spin up interact with free-layer magnetization to switch it parallel to the pinned layer magnetization (write 0), while electrons with spin down are reflected at the pinned-layer surface.

In the lower part of the figure, the case in which electrons are flowing from the right is depicted. In this case, electrons with spin up cross the whole junction and through the pinned layer without any interaction. The pinned layer acts as a reflection filter rejecting electrons with spin down that interact with free-layer magnetization to switch it anti-parallel to the pinned layer.

Due to this mechanism, MTJ can be set in a low resistance condition and reset in a high resistance condition simply by reversing the direction of the current flow.

Figure 33 shows a section of a STT-RAM cell, where one can see that a second word-line to write is no longer necessary, and the structure is much simpler and compact than the old one.

From the viewpoint of STT-MRAM performance, one can recognize that it is a very good candidate for DRAM replacement because of these characteristics:

- Compact cell (6 F^2 possible).
- High speed.
- Symmetrical read/write latency.
- Very high endurance.
- No refresh required.

The last feature in particular is a key advantage over DRAM because most of the power consumed in large DRAM banks is used for data refresh.

Also a remarkable feature for the scaling path is that the current needed to flip the magnetization of the free layer (I_c) decreases with cell scaling, unlike the old approach for achieving low-power consumption. Moreover, adoption of perpendicular (out-of-plane) magnetization enables further reduction of the cell's dimensions and a further reduction of current needed to flip the free layer (I_c), see Table 3 for a summary of different approaches.

Table 3 Summary of the characteristics of first and recent approaches to MRAM

	Conventional MRAM	STT-MRAM (in-plane magnetization)	STT-MRAM (out-of-plane magnetization)
Operation principle	Field switching	STT switching	STT switching
Characteristics	I_c increase with MTJ size	I_c decrease with MTJ size	I_c decrease with MTJ size
	Disturb in half-selected cells	Aspect ratio 2:1	Aspect ratio 1:1
Cell area (F)	>20	6–12	4–12
Scalability limits (nm)	~90	20–40	<20

Switching current I_c is defined as:

$$I_C = I_{C0}[1(K_BT/E_b)\ln(\tau_p/\tau_0)] \tag{4}$$

I_{C0} = intrinsic threshold current depending on spin transfer efficiency in the material

K_B = Boltzmann's constant

T = absolute temperature

E_b = Energy barrier to flip free layer magnetization

τ_p = write pulse width

τ_0 = natural time constant assumed 1 ns

from (4) it can be seen that I_c depends on the material, in particular on the energy barrier to flip the free layer and on the switching pulse width. It can be seen that I_{c0} and consequently I_c is different for $(0\rightarrow1)$ and $(1\rightarrow0)$. I_c determines the writing current and then the writing speed and the cell size through the size of selector element, because, if higher current is needed to write, selector transistor must be larger to be able to supply that current [9, 12].

We have already seen that the difference between high resistance and low resistance in the cell is driven by the value of TMR, but, looking at the I/V characteristic of a typical MTJ junction (Fig. 34), we can notice that this difference is also a function of the voltage applied.

A consequence of this shape is that read should be done at as low a voltage as possible to take advantage of a better read window, and also the different slopes of the two resistance curves could be used as a read method alternative to resistance detection.

The data retention in STT-RAM can be evaluated by defining the parameter Thermal Stability (Δ), which is given by:

Fig. 34 Simulation of a MTJ R/V characteristic

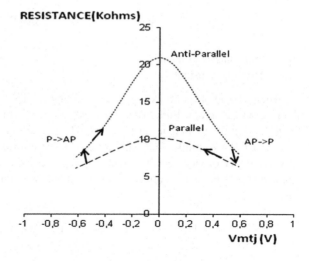

$$\Delta = E_b/K_BT \tag{5}$$

E_b is the energy required to flip the free (weak) layer. This parameter then is an indication of retention ability of the memory cell. Thermal stability is about constant with free layer volume up to a critical value, after which it is decreasing with free layer volume. Another way to write thermal stability which highlights the dependency on the material properties and layer volume (for out-of-plane magnetization) is the following:

$$\Delta = M_s VH_k/2K_BT \tag{6}$$

H_k = out-of-plane uniaxial anisotropy
E_b = energy of anisotropy
Ms = saturation magnetization
K_B = Boltzmann constant
T = absolute temperature
V = volume of the free layer

The dependency of thermal stability on the volume could raise the question of stability during the scaling path. Considering the retention requirements for a Gbit-scale product, a $\Delta > 60$–80 should be needed to guarantee an acceptable retention [13].

Actually, an attempt to increase thermal stability to have longer retention results in an increase of Ico because of the relationship shown in Eq. 7, which means more difficulty to write.

$$Ico = 2(e\alpha/h\eta) \times (2\Delta K_B T) \tag{7}$$

$\eta =$ spin transfer efficiency,
$\alpha =$ damping constant

Then a trade-off is necessary when deciding on device performance levels. One can recognize here a similarity to PCM behavior, in which, to increase retention at high temperatures, the current to melt GST must be increased.

In STT-RAM, unlike PCM and Re-RAM, the transition from high resistance to low resistance or vice versa is a switch-like process: the flip of the free layer can be treated as a statistical phenomenon in which the switching probability can be given as:

$$P_{sw} = 1 - \exp\left(-t_{pw}/t_0 \exp(-\Delta(1 - I_c/I_{c0}))\right) \tag{8}$$

$\tau_p =$ write pulse width
$\tau_0 =$ natural time constant
$I_c =$ switching current

From (8), it can be understood that the switching probability is a strong function of the critical current I_{c0}, so a small current increase over the critical value is enough to flip magnetization of the free layer.

Switching probability drives the Write Error Rate (WER) performance that is fundamental for a memory product. Following (8), to reduce WER requires longer writing pulse and/or higher writing current. Also, at a fixed switching probability, increasing write speed requires a higher writing current.

4.4   Ferroelectric Memories (FeRAM)

The concept of ferroelectric memory that has been treated in chapter "Physics and Technology of Emerging Non-Volatile Memories" is not new, and it has been used in memory products for many years, but ferroelectric memories were limited to low-density memory products [35], like SIM cards, microcontrollers, and meters, because, in the ferroelectric material used (PZT), polarization decreases with thickness of the film and below about 30 nm it is too low to ensure a stable reading of the cell. Unfortunately, 30 nm is too thick to allow high-density integration. In addition, ferroelectric material doesn't have good compatibility with the standard CMOS process which necessitates segregation of some of the process steps during fabrication.

The discovery of the ferroelectric properties of HfSiO$_x$ has completely changed the picture, opening the possibility that high-density FeRAM can compete with DRAMs.

Fig. 35 Ferroelectric
capacitor hysteresis loop

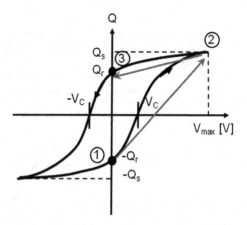

To understand the operation principle of a ferroelectric memory cell [36], it is useful to start with the hysteresis loop of a ferroelectric capacitor, which is shown in Fig. 35. Such a capacitor is simply a layer of ferroelectric material sandwiched between two conductive layers which function as electrodes.

If we draw a diagram reporting the charge Q in the capacitor as a function of the applied voltage we obtain the shape shown in Fig. 35 said capacitor hysteresis loop. When the voltage across the capacitor is 0 v, the charge in the capacitor is different from zero and the capacitor assumes one of the two stable states: "0" or "1." The total charge stored on the capacitor is Q_r for a "0" or $-Q_r$ for a "1." A "1" can be switched to a "0" by applying a positive voltage across the capacitor corresponding to the path 1–2 in Fig. 35 and then removing it (path 2–3). By doing so, the charge on the capacitor at the end of the operation will be Q_r and the total charge change will be $2Q_r$. This change of charge is the total available signal that can be detected. Similarly, a "0" can be switched back to a "1" by applying a negative voltage pulse across the capacitor, hence restoring the capacitor charge to $-Q_r$.

A FeRAM cell can be built simply by substituting the capacitor of a DRAM cell like in Fig. 36.

Fig. 36 1C-1T ferroelectric
memory cell. This structure is
very similar to a DRAM cell

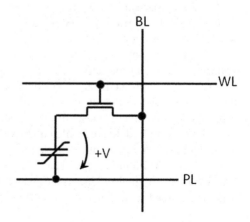

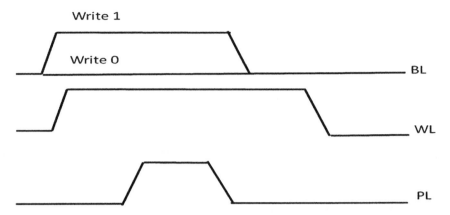

Fig. 37 Ferroelectric memory write timing. Design of FeRAM memory can borrow much of the circuit design from DRAM

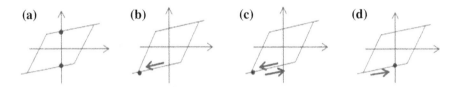

Fig. 38 Write 1 operation path along the hysteresis characteristic

As in DRAM, the MOS transistor called the access transistor controls the access to the capacitor and prevents any disturbance. When the access transistor is off, the ferroelectric (FE) capacitor remains disconnected from bitline (BL) and hence cannot be disturbed. When the access transistor is ON, the FE capacitor is connected to the bit-line and can be written or read by the plate-line (PL).

The cell is accessed by raising the word-line (WL) and hence turning ON the access transistor. The access is one of two types: a write access or a read access.

The timing diagram for a write operation is shown in Fig. 37.

In Fig. 38 the writing process of a cell is shown on the hysteresis curve of ferroelectric capacitor.

To write a "1" into the memory cell, initially the BL is raised to Vdd (negative voltage across the ferroelectric capacitor following the convention of Fig. 36); at a later time, the word-line is raised to Vdd + V_t, where V_t is the threshold voltage of the access transistor, this enables the full Vdd to appear across the ferroelectric capacitor. At this time, the state of the ferroelectric capacitor is shown in Fig. 38b, then, the PL is pulsed, that is, pulled up to Vdd and subsequently pulled back down to ground, which represents a back and forth movement along the hysteresis characteristic shown in Fig. 38c. The final state of the capacitor is a negative-charge state (defined as digital "1" in this chapter). Finally, deactivating the WL leaves this state undisturbed until the next access (Fig. 38d).

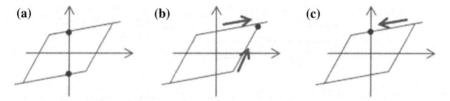

Fig. 39 Sequence of writing a "0" as seen on the hysteresis curve

To write a "0" into the cell, (Fig. 39) the BL is driven to 0 V prior to activating the WL. Then, the line PL is pulled to Vdd. This time a positive voltage appears across the ferroelectric capacitor which is in the position indicated in Fig. 39b. When PL comes back to zero, the state of the ferroelectric capacitor has changed as in Fig. 39c regardless of its initial state (Fig. 39a).

Comparing the sequence of states depicted in Figs. 38 and 39, we realize that pulsing PL during a "1" writing is not really required, and raising BL to Vdd is actually enough to complete the operation. In fact, PL pulsing produces a simple back and forth movement of ferroelectric capacitor state as indicated in Fig. 38c. The choice to give a pulse to PL is done because in this way PL can be a common line in the array, regardless if a bit should be written to "1" or to "0".

To detect the status of ferroelectric capacitor, or in other words the direction of polarization, it is necessary to apply a voltage $> V_c$ across the capacitor. Referring again to Fig. 35, if the capacitor was initially at point 1 of the characteristic, the voltage pulse applied corresponds to a transition from (1) to (2) and return to (3) moving a charge $2Q_r$ determined by the polarization switch. But, if the capacitor was originally in (3), the path is from (3) to (2) and back to (3), moving only the charge $Q_s - Q_r$ coming from displacement component of the capacitor. Signal detection in a ferroelectric cell is a destructive operation, and the original data must be restored in the cell exactly as it happens with DRAM.

The cell we have seen so far, called 1T-1C (1 transistor-1capacitor), is not the only possibility to use a ferroelectric layer to build a memory cell.

Another device that can be conceived, called FeFET [37], is simply a MOS transistor in which the gate oxide is replaced with a ferroelectric layer (Fig. 40). Polarization of the ferroelectric layer is able to make the threshold voltage positive or negative, depending on the dielectric polarization direction. As in a flash cell, data will be read by detecting the threshold shift.

The FeFET transistor was proposed to create a FeNAND memory cell (Fig. 41) that has an architecture similar to NAND and operates in the same manner as normal NAND, with some advantages: high cycling capability in the order of 10^8 $^-10^9$ (on single cells but still not demonstrated on large arrays), <5 V operation, a potential advantage in scaling. In the competition with NAND, FeNAND retention has an issue due to FeFET depolarization effects, and MLC seems very difficult to implement due to the small read window. Investigations are ongoing about the possibility of replacing NAND cells with FeNAND in any of 3D NAND flows.

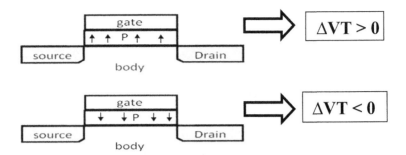

Fig. 40 FeFET device: direction of polarization P in the ferroelectric layer under the gate determines a threshold voltage shift in one direction or the other, which can be detected as in the well-known flash memory cell

Fig. 41 FeNAND cell

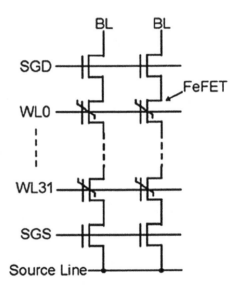

FeRAM as a DRAM replacement enables high-speed read and write, together with low-power operation. As in DRAM, a displacement current is involved, and writes energy per bit is similar to DRAM, about 2 pJ/bit.

If we consider a 1T-1C cell, which is the most popular configuration, endurance is on the order of $>10^{13}$ cycles, which is somewhat less than DRAM, but probably still acceptable.

The main problem of FeRAM regarding endurance is a phenomenon called "fatigue". It may be defined as a reduction of the residual polarization with cycling. Above a certain level, the signal available for reading is too small to allow correct operation (see Fig. 42).

Fatigue is correlated to a re-ordering of material microscopic structure involving oxygen vacancy.

Fig. 42 Polarization loss
upon cycling (fatigue)

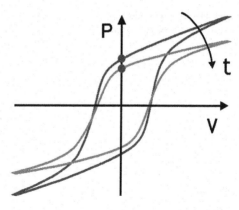

Fig. 43 Shift of the
hysteresis loop (imprint)

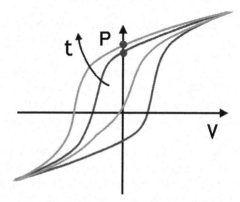

Endurance can also be affected by "imprinting" or "aging" which is a polarization on one of the two possible states which makes very difficult to change the status of the material after cycling (see Fig. 43).

This effect is caused by a distortion of the ferroelectric domains due to the capture of charged particles like electrons or charged species.

As mentioned with regard to FeFETs, retention can be affected by depolarization which is a loss of polarization over time (see Fig. 44) depending on the initial relaxation after programming.

4.5   Final Considerations

From the previous discussion on memory technologies under development, it becomes clear that the goal of realizing a universal memory capable of replacing volatile and non-volatile mainstream technologies (DRAM and NAND) is not easy to realize, at least in the short term. Instead, we can identify a group of technologies more suitable to cover NAND performances and others to cover DRAM products.

Fig. 44 Time-dependent
polarization loss (retention)

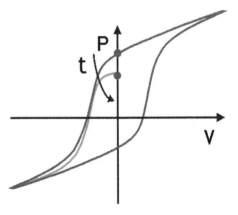

To be a credible alternative to NAND it is necessary to match its fabrication cost
and be able to produce the multi-GBit density scales which imply the possibility to
have 3D and MLC technology with a good scaling perspective and at reduced cost.

Among the technologies we have previously reviewed, FeRAM and
STT-MRAM don't have MLC capability; STT-MRAM is a very complex technology; and 3D, even if it in principle is feasible, is very difficult to realize.
FeRAM, as we have seen is a simpler technology but also for it 3D is very difficult
to realize.

On the contrary, PCM and ReRAM have at least in principle the capability for
MLC and, provided that a suitable selector device is integrated, they can be used for
a 3D approach, and the technology is relatively simple and scalable, so they could
be good candidates for NAND replacement, having in addition better scaling perspective and bit-level granularity. An advantage of ReRAM is a lower write current
and energy consumption and potentially an easier MLC implementation.

PCM instead, due to a higher maturity level, can show consolidated data on large
arrays, which are more indicative of the real capability of the technology.

Set/Reset times are faster for ReRAM with respect to PCM, but it has to be
considered that it is not possible to use a single pulse to write the cells because the
current (or resistance) distribution obtained in this way would be too large, and
correct operation would be impossible. In general, a sequence of pulses with various durations and heights with a verify operation after each pulse will be necessary
to obtain narrow distributions and consequently correct and reliable operation of the
device.

The maturity levels of the technologies we have described so far vary significantly from one to the other. Considering the progress done in integration of large
arrays, we can see that emerging memories are still well below NAND, as far as
memory density is concerned, by at least of a factor of 4 (see Fig. 45). Currently
only PCM, FeRAM, and MRAM are in production, and PCM has an order of
magnitude lead over the other emerging memories. Recent acceleration in the study

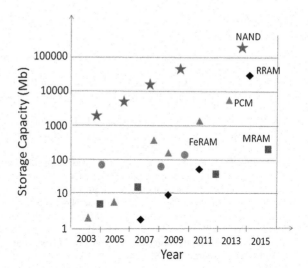

Fig. 45 Storage capacity development of some of emerging memories

of 3D PCM and FeRAM can affect a change of slope to the speed of development of high-density memories using these two approaches.

Read-latency time on the order of 50 ns or even faster, is reported for all the technologies competing with mainstream approach, even if the signal available varies much between the various different technologies. Except for FeRAM that is based on capacitance share, the ratio Ion/Ioff of the other memories can be taken as a figure of signal available for reading: the higher the signal, the easier and faster should be reading. It has been found that this parameter has a large variation over the different technologies: it can span from 1000 in PCM to 2 in STT-RAM.

Technologies like FeRAM and STT-MRAM exhibit almost symmetrical SET and RESET times, on the order of 10–50 ns and comparable with read time, in addition to endurance on the order of 10^{13}. In addition, they are intrinsically non-volatile memories, and, for this reason, they have been considered for DRAM replacement as non-volatility enables avoiding the power-consuming refresh operation which is one of the blocking points for the growth of big data centers. On the other hand, a price would be paid in terms of a 20–30% degradation of cycle time with respect to DRAM, so a full DRAM replacement would be feasible for slower applications.

STT-RAM has a good performance in endurance, but, on the other hand, the signal available is poor and probably, a high Bit Error Rate will lead to a mandatory Error Correction Code (ECC). In addition, high current is required for write (switch), and compatibility with existing processes is poor. If we compare both the technologies with DRAM, we find about +20% read latency paid by STT-RAM due to the poor signal available, while FeRAM pays less than 10%. Another good feature for FeRAM is that ECC is likely not required. ECC not only consumes silicon area, but also results in a further slowdown of read speed.

References

1. J S Meena et Al. "Overview of non-volatile memory technologies" Nanoscale Research Letters 2014, 9:526
2. H. Tanaka et Al. "Bit Cost Scalabl Technology with Punch and Plug Process for Ultra High density Flash memory" Proceedings of 2007 IEEE Symposium on VLSI technology, pp 14-15
3. J.Jang et Al. "Vertical cell array using TCAT (Terabit Cell Array Transistor) technology for ultra-high density NAND Flash memory" Proceedings of 2009 Symposium on VLSI technology pp. 192-193
4. R. Katsumata et Al. "Pipe –shaped BiCS flash memory with 16 stacked layers and multi-level-cell operation for ultra-high density storage devices" Proceedings of 2009 IEEE Symposium on VLSI technology pp. 136–137
5. W. Kim et Al. "Multilayered Vertical Gate Nand Flash overcoming stacking limit for terabit density storage" Proceedings of 2009 IEEE Symposium on VLSI technology, pp.188-189.
6. J.Kim et Al. "Novel Vertical-Stacked-Array-Transistor (VSAT) for ultra-high density and cost effective NAND flash memory devices and SSD (solid state drive)" Proceedings of the IEEE Symposium on VLSI technology, pp.186-187
7. K.Kim et Al."Future outlook of NAND Flash technology for 40nm Node and beyond" Proceedings of the IEEE Non-Volatile Semiconductor Memory Workshop, Monterey CA, 2006, pp.9-11
8. ISSCC2009 Forum "SSD Memory Subsytem Innovation" chair: K.Takeuchi
9. R.F. Freitas et Al. "Storage Class Memory: the next storage system technology" IBM Journal of Research and Development, 2008.
10. G.W.Burr et Al. "Overview of candidate device technologies for storage class memory" IBM Journal of Research and Development vol.32, No 4/5 July/Sept. 2008.
11. D.Adler et Al. "The mechanism of threshold switching in amorphous alloys" Review of modern physics, vol.50, n.2, April 1978
12. S.R.Ovshinsky "Amorphous materials as optical information media" Journal of Applied Photographic Engineering Vol.3 No.1, pp.35-39 (Winter, 1977)
13. S.R.Ovshinsky et Al. "Amorphous semiconductors for switching, memory and imaging applications" IEEE Transactions on Electron Devices vol.20, issue:2, Feb.1973.
14. G.W.Burr et Al. "Phase change memory technology" IBM Journal of Research and Development, 2009
15. D.Ielmini et Al. "Analytical model for subthreshold conduction and threshold switching in chalcogenide-based memory devices" Journal of Applied Physics, 102(5), Sept.2005
16. A.Pirovano et Al. "Scaling Analysis of Phase-Change memory technology" IEDM2003 Technical Digest pp. 29.6.1-29.6.4
17. G.H.Koh "Emerging memory technology" 2012 VLSI short course
18. J:J:Yang et Al. "Memristive switching mechanism for metal/oxide/metal nanodevices" Nature Nanotechnology Vol.3, Jul. 2008
19. M.N: Kozicki et Al. "Information storage using nanoscale electrodeposition on metal in solid electrolyte" Superlattices and Microstructures 34 (2003) 459-465
20. H.S. Wong et Al. "Metal Oxide RRAM" invited paper, Proceedings of the IEEE, Vol.100, No. 6, pp.1951-1970, June 2012
21. H.Akinaga et Al. "Resistive Random Access Memory (ReRAM) based on metal oxides" Proceedings of the IEEE No. 12, Dec.2010.
22. Chakravarthy Gopalan, Yi Ma, Tony Gallo, Janet Wang, Ed Runnion, Juan Saenz, Foroozan Koushan, and Shane Hollmer "Demonstration of Conductive Bridge Random Access Memory (CBRAM) in logic CMOS process
23. S.T.Hsu et Al. "RRAM switching mechanism" 2005

24. S.Yu et Al. "On the switching parameter variation of Metal Oxide RRAM - part II: model corroboration and Device design strategy" IEEE Transaction on Electron Devices, Vol.59, No.5 pp.1182-1188,2012
25. X. Guan, S. Yu, H.-S. P. Wong, "A SPICE Compact Model of Metal Oxide Resistive Switching Memory with Variations," IEEE Electron Device Letters, vol. 33, No.10, pp. 1405 – 1407, October 2012
26. X.Guan et Al. "On the Switching Parameter Variation of Metal Oxide RRAM - part.I: Physical modeling and Simulation Methodology" IEEE Transaction on Electron Devices, vol.59, No.5, pp.1172-1182,2012
27. Wong "Emerging Memory Devices" in: ISDR Symposium, College Park, Piscataway, Dec. 2011
28. B.Wang "Emerging Technology Analysis: the future and opportunities for next generation memory" Gartner Inc: Stanford, 2011
29. W.J: Gallagher et Al. "Development of the Magnetic Tunnel Junction MRAM at IBM: from first functions to a 16Mb MRAM demonstrator chip" IBM Journal of Research and Development, 2006, 50 (1); 5-23
30. M.Julliere "Tunneling between ferromagnetic films" Physics Letters A, 1975, 54(3);225-226
31. L. Berger "Emission of spin waves by a magnetic multilayer traversed by a current" Phys. Rev. B 1996; 54(13):9353–8
32. JC. Slonczewski "Current-driven excitation of magnetis multilayers" IEEE Journal of Magnetic Materials 1996; 1859(1/2):L1–7
33. JC.Slonczewiski "Currents, torques and polarization factors in magnetic tunnel junctions" Phys. Rev. B, 2005;75(10)
34. T.Kawahara et Al. "Spin-transfer torque RAM technology: review and prospect" Microelectronics Reliability, 2012(52), pp. 613-627
35. S.Masui et Al. "FeRAM application for next generation smart card LSIs" in Proc. 1st Int. Meeting Ferroelectric Random Access Memories, 2001; pp.13-14
36. S.Chung "circuits and Design Challenges for Alternative and Emerging NVM" in: Non Volatile Memory Design Forum, ISSCC, Feb.2007
37. X.Liu et Al. "Ferroelectric memory based on nanostructure" Nanoscale Res. Lett. 2012;7 (1):285

Array Organization in Emerging Memories

Roberto Gastaldi

1 Introduction

Memory cells need to be packed together in a memory array to be used. Arrays are generally made of rows and columns, and a single cell in the array is perfectly defined as the crossing point of a particular row with a particular column; then, in some way, one single column and one single row of the array have to be connected to a particular memory cell. In general, to avoid mutual disturbances or unwanted data modifications, a single memory cell is connected to its column and row through a "selector" device. Everybody has in mind the case of flash cell, where the selector and memory element are merged within a single device. In that case, the memory effect is characterized by a change in an MOS-transistor threshold due to negative charge storage in the so called "floating gate". The gate of a flash cell is connected to a common word-line, while the drain is connected to a common bit-line. When a word-line is switched to ground, all the cells connected to that word-line are disconnected from the array. In all emerging memories, on the contrary, the selector device and memory material are separated, and, generally, fabrication of memory material occurs in the process back-end. This opens the possibility to make different choices for the selector element, optimizing its characteristics with respect to the conditions required by the particular material used. Moreover, most of the emerging memory devices operate in a bipolar mode, which means that, to change the state of memory cell, a reversal of applied voltage is required, and so the selector transfer characteristic must allow a transfer voltage in both directions with similar efficiency.

R. Gastaldi (✉)
Redcat Devices s.r.l, Milano, Italy
e-mail: r.gastaldi@redcatdevices.it

© Springer International Publishing AG 2017   89
R. Gastaldi and G. Campardo (eds.), *In Search of the Next Memory*,
DOI 10.1007/978-3-319-47724-4_5

2 PCM and Array Organization for Unipolar Operation

We have seen in the previous chapter that PCM cell is an unipolar device (changing the state of the cell doesn't require applied voltage reversal) and it can be taken as an example to explain unipolar array organization.

The simpler way to organize the array of PCM is to use a NMOS transistor as a selector and connect the cells in a NOR configuration, as in Fig. 1, where the memory element is schematized with a (variable) resistance.

This configuration has the advantage to be very well-known, and we can proceed further in the design more or less as done in an flash-NOR array.

Stand-by current is an important parameter for memories, and, in multi Gbits memories, the leakage current from unselected cells in the array is an important contribution to the total stand-by current, and it is carefully evaluated during design. An NMOS selector offers a very low I_{off} (the leakage current when the cell is deselected) provided that a non-minimum length device is designed and, thanks to its very high input impedance, doesn't load row decoders with DC current. MOS transistor can also be driven in the saturation region during write operation mode to control the set/reset current.

Together with the advantages mentioned above, an MOS-transistor selector brings some drawbacks: first of all, a large area occupation, which is a limitation for larger memory size, in particular if a planar MOS is used; and the width of the transistor cannot be too small because it must be able to provide the necessary programming current, that, in the PCM case, can be relatively large (on the order of 200–500 μA). This constraint is worsened by the consideration that to maintain a low-enough I_{off}, the choice of a minimum-length device should be avoided. If the stand-by current budget for an array is I_{sb} the limit for leakage current of a single cell is:

Fig. 1 NOR/type array organization of PCM cells

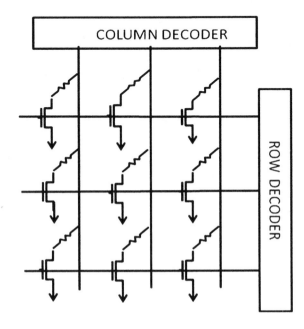

$$I_{sbcell} = I_{sb}/N \qquad (1)$$

where N is the total number of cells in the array.

Cell leakage may become a problem, not only during the stand-by condition as discussed above, but also during reading: for a long bit-line, the leakage of non-selected cells connected to the same bit-line may be important and create a disturbance when reading a zero (corresponding to low current in the cell). A design condition to be met to avoid reading a "1" instead of "0" due to leakage is:

$$I_{leakTot} = (N-1)I_{leak} < I_{"0"max}/10 \qquad (2)$$

Which means that the total leakage on a bit-line must be ten times lower than the maximum current of a cell in a "0" distribution.

A possibility to achieve a more compact cell is splitting the selector layout to have two fingers that give the same width but consume less area; however, a cell with an MOS selector occupies a cell area of about 15–20 F^2, which is too big for high-density memory devices.

Biasing of the array is quite similar to the well-known NOR-flash array: all the non-selected word-lines and bit-lines are at ground, while the selected word-line is brought to positive voltage, having only a capacitive load. In a similar way, the selected bit-line is raised and the read, set, and reset voltages are passed through it.

A much better solution to reduce the global area occupation of the cell is to use a bipolar device (BJT or P-N diode) as a selector, because it can be fabricated in the silicon region directly beneath the memory stack (Fig. 2).

Fig. 2 A bipolar selector can be fabricated beneath the memory stack

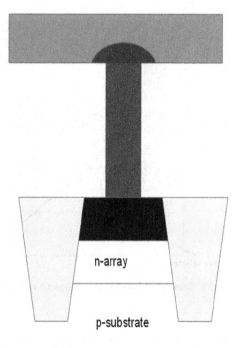

You can see from the figure that the selector is in this case a PNP transistor with an emitter connected to the heater of the GST stack and a collector connected to the common substrate ground, while the base must be connected to a metal line connecting together all the cells lying on a word-line. This configuration leads to an 8–10 F^2 cell size, then it is much more viable for large memory sizes. In addition, a bipolar device is able to sustain higher current in write.

A similar solution, yet more compact, is to adopt a true vertical p-n diode as a selector, which leads to 6 F^2 cell.

When BJT selected cells are assembled in an array, those cells sharing the same word-line (the same N type base of selector device) are affected by a parasitic lateral PNP device, that take place between adjacent P^+ areas and the common N-type layer, disturbing neighboring unselected cells near a selected one as schematically shown in Fig. 3.

It is then very important to cut the gain of parasitic PNP, for example increasing the distance among the cells, as the efficiency of a parasitic PNP rapidly decreases with distance, or separating each cell from the other. Nevertheless, since these solutions increase the cell area, in most recent technology the solution adopted is to increase the doping of the N-base region which also allows decreasing the number of contacts to metal line to have more compact cells.

Using a BJT device as a selector produces a big impact on the array organization because now word-lines and bit-lines are linked with a P-N junction that can allow a leakage current flowing from bit-line to word-line through the base of the bipolar device. The situation is schematized in Fig. 4. It should be noted that, due to the fact that selector is a PNP device whose base terminal is connected to word-lines, unselected cells must have positive voltage applied to the word-lines, while selected

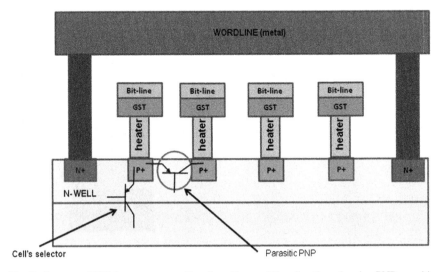

Fig. 3 Conceptual PCM-array cross-section along the word-line direction, showing PNP parasitic issue

Fig. 4 Leakage currents (in *red*) and program current (in *blue*) in a bipolar selected array

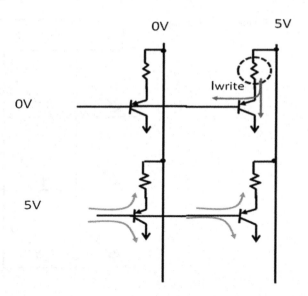

cell will have its word-line at ground; thus the logic of row decoder is inverted with respect to the familiar MOS array.

In Fig. 4, the program configuration is shown and the selected cell is identified by a dashed circle. The programming current is actually the emitter current (I_e) of the BJT selector, and it is shared between the collector and base of the device depending on current gain β according to the relationship:

$$I_e = I_b + I_c \quad \text{and} \quad I_c = \beta I_b \tag{3}$$

From the previous relationship, we can find:

$$I_e = (\beta + 1)I_b \quad \text{or} \quad I_b = I_e/(\beta + 1) \tag{4}$$

From the last relationship, we can understand that the fraction of emitter current flowing into the base is higher the lower is the current gain β of the transistor, and, in the limit of very low β emitter and base, currents will be almost equal, or the BJT behaves like a P-N diode. In a practical realization of an array, β is very low for technology reasons and also because a high β, that could bring a benefit reducing the current flowing into the selected word-line, would result in a more resistive word-line itself which counterbalances the effect of current reduction and causes a worsening of parasitic effects. In conclusion, we can assume that a big part of the emitter current flows into the selected word-line, and, for this reason, in some of the analysis that we will present in the following section, the array will be modeled as if the selector were a diode. By in some practical realizations, the selector of the PCM cell is a real diode as shown in Fig. 5.

Fig. 5 Diode selected array

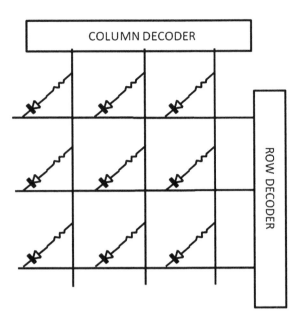

An important issue comes from the unselected cells in the array. As can be seen from Fig. 4, the transistors of these cells have two kinds of leakage: a current through the base-emitter junction and one through the base-collector junction (both reverse biased) even if the former is the most important contribution. The sum of these contributions in an array of n rows by m columns is given by:

$$I_{leakage} = \Sigma_{i=1,n}(\Sigma_{j=1,m-1}(I_{be})_{i,j} + \Sigma_{j=1,m}(I_{bc})_{i,j}) \tag{5}$$

The contribution of leakage current to the total power consumption allowed in stand-by conditions for large arrays is very important and may be a serious barrier with respect to the assigned stand-by specification for the memory product.

For example, if the stand-by budget current assigned to the array is Isb = 50 μA and the array is made of 1 Gbits, a single bit's leakage current should be less than 50 fA at maximum operating temperature. To overcome this issue, sometimes it is necessary to partition the array into a number of sub-arrays (often called tiles) for which the leakage specification is sustainable. Only one of the sub-arrays is biased in the active mode, while the others are left in sleeping mode. When a read operation is initiated, the read latency depends on the position of the cell addressed: if it is inside the activated sub-array, read latency will correspond to a normal access from the stand-by condition, otherwise the current sub-array is de-activated, and the right sub-array must be awakened, raising all the word-lines to high voltage except the one that is selected. This operation will require additional time which leads to an increased latency. The choice of the size of a single tile is generally the largest possible, compatibly with the leakage per tile value, the voltage drop on word-lines

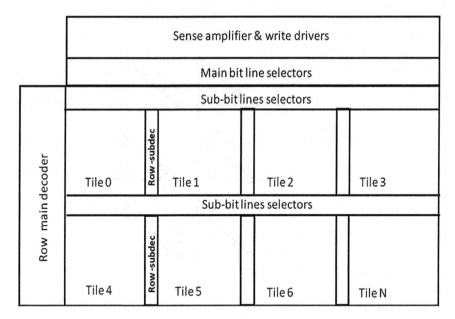

Fig. 6 Conceptual drawing of array partitioning architecture

and bit-lines dependent on the resistance of the material and the delay introduced by the resistance and capacitance of the lines. The concept of array partitioning is showed in the drawing of Fig. 6: the array is divided in tiles that can be individually selected by column and row sub-decoders driven by main decoding system. In real devices partitioning of the array is done with a hierarchical architecture grouping the tiles in larger blocks.

The presence of the leakage issue leads to relevant variations in the decoding scheme for memories using BJT/P-N diode selectors; in fact, unselected bit-lines cannot be put to ground because leakage would be too high, but, on the other hand, they cannot also be left floating as we can understand referring to the schematic array of Fig. 7, in which we have assumed that the selector is an ideal bipolar diode to simplify the schematic. In Fig. 7, if we consider the polarization scheme in which the non-selected bit-lines are at ground, then the full Vr will be applied to the cells at crossing points of non-selected rows and non-selected columns, generating a very high total leakage in the array which certainly will not be compatible with stand-by current specifications of the device. In Fig. 8, the non-selected bit-lines are left floating, but unfortunately this creates a spurious current path that charges up the non-selected bit-lines to an intermediate voltage depending on the resistance of the cells. The resulting leakage current could still be too high to be allowed and in addition the cells connected to the selected row and un-selected columns receive the leakage currents of all the cells in the column and a spurious write can result. Finally, Fig. 9 shows the case in which un-selected bit-lines are biased at a small

Fig. 7 Leakage effect on grounded, non-selected *bit-lines* in a diode selected array

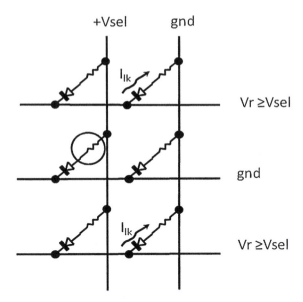

Fig. 8 Leakage effect in floating, non-selected *bit-lines* in a diode selected array

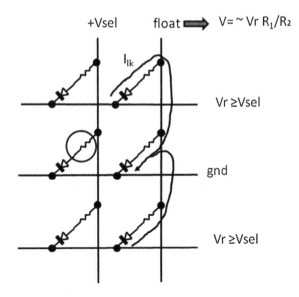

voltage, allowing a reduction of leakage to acceptable value and avoiding cells disturbances.

In the array of Fig. 9 the selected cell is identified by a circle, the corresponding bit-line is biased at a voltage $+V_{sel}$ while unselected bit-lines are maintained at about 0.3–0.4 V in such a way as to decrease the reverse biasing on diode and cut the leakage current. The bias of the array shown in the figure corresponds to the

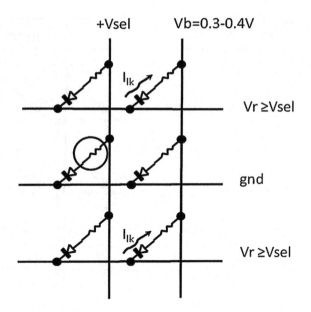

Fig. 9 Minimizing leakage on non-selected *bit-line* in a diode selected array

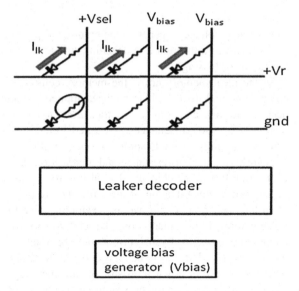

Fig. 10 Principle schematic of the "leaker" bias of a diode-selected array

reset condition because this involves the highest voltages and so the higher reverse biasing. During read, the situation is similar but much less severe because the bias is much lower.

In Fig. 10, a principle schematic of a bias circuit for unselected bit-lines is depicted [1]. The small bias voltage on unselected bit-lines may be obtained with a dedicated circuit, for example including a MOS transistor for each bit-line driven by

Fig. 11 Equivalent circuit of
a diode-selected array
including parasitic elements

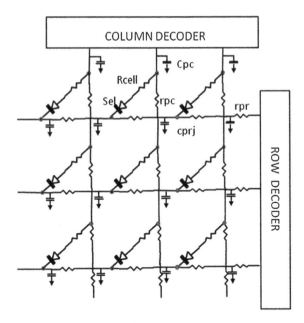

suitable gate voltage to generate the required V_{DS} when the total leakage current of the bit-line flows through it, a decoding structure is needed to exclude the selected bit-line from biasing.

In the ideal memory array, the memory cells are connected to a net of perfect conductors, and then the bias voltages applied through row decoders and column decoders to rows and columns are transferred to each cell without variations. In this ideal picture also capacitances associated to the connecting lines in the array are zero, so the pulsed signals applied propagate along the lines without delay. In the real world, a parasitic resistance and a parasitic capacitance is associated with any connection line in the chip and then also to rows and columns of memory array, so that a correct modellization of the array should appear as in Fig. 11.

In the array architecture we are examining, due to these parasitic resistances a voltage drop will take place along the selected word-line, which is dependent on the magnitude of the program current and the total resistance associated with the word-line itself. This has to be considered in the design of the decoding circuitry to dimension the pull-down transistor of the final stage of the row decoder and to choose the right value for programming voltage. For example, let's consider the circuit in Fig. 12 that is a modelisation of a bit-line during reset in a PCM array. We consider the case of reset because this operation requires the higher current, but the situation is similar for set or read condition, even if the voltages are lower in this case and the drawbacks are smaller.

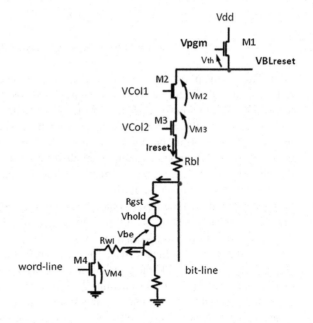

Fig. 12 Simplified schematic of current path through a selected cell in a bipolar-selected memory array

In this figure, M_1 controls the voltage applied to the bit-line, M_2, M_3 represent column selectors hierarchy and R_{bl} and R_{wl} are the parasitic resistances of bit-line and word-line while the memory element is schematized with a (variable) resistance R_{gst} and a voltage generator equal to V_{hold}, V_{be} is the base-emitter voltage of the BJT selector and finally M_4 represents the pull-down transistor of final stage of row decoder The voltage applied to the bit-line during reset (VBL_{reset}) is equal to $V_{pgm} - V_{th(M1)}$ and its value is given by (6):

$$VBL_{reset} = 2V_{M2,3} + V_{hold} + R_{gst}I_{reset} + Vbe + R_{bl}I_{reset} + \Delta V_{reset} \qquad (6)$$

We can understand that the value to be chosen for VBL_{reset} (and then for V_{pgm}) must take into account the voltage drop on column selectors and the effect of bit-line resistance, but due to the BJT selector an additional ΔV_{reset} must be added to the budget. This additional contribution is caused by the current flowing out of the base of BJT through the word-line and the transitor M_4 The current flowing out from the base of selector transistor is dependent from its current gain β, as we know from (4). Then a higher β can reduce ΔV_{reset}. If some cells lying on this word-line are programmed in parallel, the pull down of the final stage of the row decoder must be dimensioned to sink the reset current of all these cells. As the reset current is in the range of 500 µA/cell, this condition sets a strict limitation on the number of cells sharing the same word-line that can be programmed in parallel. Another issue is that the voltage drop caused by parasitic elements reduces the current capability of BJT selector reducing V_{be}.

The additional voltage ΔV_{reset} is given by:

$$\Delta V_{reset} = R_{on}nI_{reset}/(\beta+1) + \Sigma_{j=1,n}R_{WLj}jI_{reset}/(\beta+1) \tag{7}$$

where R_{WLj} is the portion of word-line parasitic resistance between the cell j and the cell j + 1, n is the total number of cells under programming in a single wordline, R_{on} is the drain to source resistance of the row-decoder pull down (M_4).

Unfortunately, the parasitic resistance seen by a single cell in the array is a function of its position, but the calculation of the required write voltage must be done to ensure that enough voltage is applied to the worst case cell. The result is that almost no cell receives the optimum writing voltage, unless some tracking and compensation system can be used.

To have an idea of the Reset voltage required in a real circuit, let's refer again to Fig. 12: if we assume $I_{reset} = 300$ μA, $V_{DS}(I_{reset})$ of all MOS transistors is about 0.1 V, $V_{be} = 1$ V, $\beta \sim 1$, $V_{HOLD} = 0.5$ V, $R_{gst}I_{reset} = 1.5$ V, and $R_{bl} = 3$ kΩ and $\Delta V_{reset} = 0.3$ V the resulting VBL_{reset} at the top of column decoders is about 4.4 v. To have V_{pgm} we must add the threshold voltage of M_1, which actually is a function of I_{reset}: let's put $V_{th} = 0.7$ V, resulting V_{pgm} is about 5.1 V. This voltage must be generated internally in the chip, starting from a supply voltage of 1.8 V. On one hand, higher voltage means more power consumption and a more demanding and area-consuming charge pump, while, on the other hand, there is the risk of breakdown of some cells requiring much less voltage. One possible action to lower programming voltage is to reduce, as much as possible, the parasitic contributions from the topological elements of the array (R_{bl}, R_{wl}) which can be accomplished with a suitable partitioning of the array. On the side of technology, efforts are devoted to reduce I_{reset}, which also has the advantage of reducing the power needed for programming. The idea of increasing the β of the bjt selector to reduce the fraction of I_{reset} flowing into the base is not feasible due to the topology of bjt and because this would worsen the disturbance caused by lateral PNP on both neighboring cells.

Another consideration should be made about column selectors (two levels of decoding are assumed): to reduce appropriately their $V_{DS,}$ which enlarge the factors increasing VBL_{reset}, pumping of their gates over V_{dd} may be required, but this solution leads to additional power consumption and adds a further complication in the design, plus consuming additional area. Another possibility is to adopt a p-channel decoding structure, since even in this case increased area is needed because of n-well of p-channel devices, but it seems less expensive than N-channel decoding.

In addition to resistances, the presence of parasitic capacitors creates unwanted effects during dynamic operation of the array. The most obvious effect is a delay in signal propagation because the chain of resistances and capacitances act as a delay line; this is important in read mode because the specifications for read latency are usually very tight, but it can cause problems also during set and reset operations which are much slower than read for example during the "quench" phase of reset pulse.

3   Bipolar-Operation Array Organization

We have seen that some of emerging technologies [2], such as Re-RAM or
STT-MRAM, require a reverse of the voltage applied to the memory stack to
perform the set and reset operations and due to this fact that the array architecture
must be designed to allow voltage reversal. This reflects immediately on the
characteristics of the selector device that must allow current flowing in both forward
and reverse directions. The bipolar diode or the BJT, which has been used as a
selector for PCM array, cannot be used in this way and must be abandoned. On the
contrary, an MOS transistor is a symmetrical-operation device and can be used for
this application. The usual NOR configuration can still be maintained (see Fig. 13),
but the common terminal (now said PLATE) cannot be put to ground due to the
need for bipolar operation, so it should have the capability to switch to write
voltage.

In the actual architecture, however the plate should be kept at a fixed voltage due
to the large capacitance associated with it that makes any voltage change very slow
and power consuming. Let's take the example of an STT-MRAM cell: in this case.
MTJ is modeled with a variable resistance and let's suppose that the required

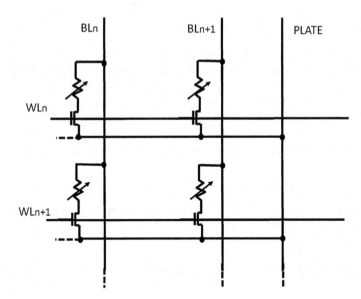

Fig. 13 Common plate array organization

voltage applied to the junction to write (set) a "1" is $V_{set} = 0.35$ V, with the positive side toward the bit-line. The voltage to be applied to the bit-line is:

$$Vbl_{set} = 0.35 + R_{on}I_{set} + V_{plate} \quad \text{(SET operation)} \tag{8}$$

where R_{on} is the selector drain resistance, I_{set} is the current needed to set the cell, and V_{plate} is a fixed voltage applied to the plate terminal. Assuming that the selector device operates in linear zone, R_{on} is given by the simplified relationship:

$$R_{on} = 1/(V_{gs} - V_t)\mu C_{ox}(W/L) \tag{9}$$

V_{ds} should be minimized, which, for a fixed technology and a given I_{set}, can be done by tuning the W/L ratio and V_{gs}. In a similar way, for RESET, if we assume for simplicity $V_{set} = V_{reset}$, $I_{set} \sim I_{reset}$ and $Vbl_{reset} = 0$, we can write

$$V_{plate} = R_{on}I_{reset} + 0.35 \quad \text{(RESET operation)} \tag{10}$$

Then, if we assume $R_{on}I_{reset} = 0.3$ V in our example, we obtain $V_{plate} = V_{reset} = 0.65$ V and

$$\begin{aligned} Vbl_{set} = V_{set} = 1.3\,\text{V} \\ Vbl_{set} \sim 2\,V_{plate} \end{aligned} \tag{11}$$

In the previous example, the effect of parasitic resistance of bit-lines and plate line has been disregarded. Also the body effect on MOS threshold determined by the bias on source has not been considered. In the real case, the situation is not the same for SET and RESET (see Fig. 14), in the first case the source of selector transistor is at a fixed voltage $V_{source} = V_{plate}$, while in the second case, V_{source} depends on the memory resistance by the reset current V_{mem} and the transconductance of MOS is modulated by the body effect. Gate voltage applied to the

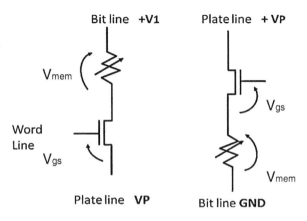

Fig. 14 Resistive memory cell operated in bipolar mode. When the source of selector is not tied to Plate, the body effect and voltage drop on memory resistance require a higher word-line voltage be used

selector must be calculated taking into account the overdrive needed to compensate this effect.

So far, we have assumed the selector in the linear operating zone reduces V_{DS} as much as possible, but if the available voltage is enough, we can imagine operating the MOS in saturation, thus controlling the current flowing into the memory stack. This is a method used to control the current through the memory stack during write, and it can be used to avoid stressing the cells, particularly, when a write algorithm is required, as in Re-RAMs and PCMs.

Even if the plate is normally at a fixed voltage during write operation and may stay at that level even in read mode, it must be charged at the beginning of active mode, and this can lead to a big increase in latency at power-on; although a solution can be to choose ground level for the plate, this leads to the need for negative voltage management in the bit-line decoding and in the array, which increases design complexity, so it can be more convenient to selectively bias the plates of the active tiles.

An architecture solution that allows independent choices of V_{set} and V_{reset} is shown in Fig. 15, where both the bit-lines and "source" lines run in parallel. Both of them must be decoded, and column decoding may be shared between top and bottom of the array [3], where selectors placed at the top of the array are dedicated to bit-lines, while those at the bottom are dedicated to source lines, and the cells are connected between these two lines. Writing "1" or "0" is obtained by alternatively driving to ground the bit-line or the source, while the other line is set at the write voltage. In spite of the fact that this solution is simple, it may result in large area

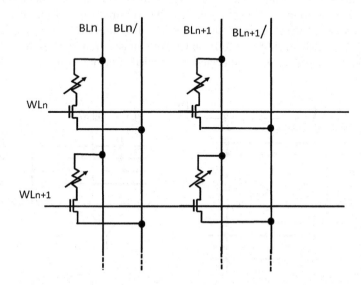

Fig. 15 BL-BL/organized array

occupation, but, on the other hand, this is the configuration that ensures minimum voltage stress and better overdrive on the selector transistor of the cell.

Considering an architecture as in Fig. 15, alternating BL and BL/, an important issue to take into account is the parasitic resistance of source lines that affects the current flowing into the cell, thus reducing the overdrive on the selector transistor of the cell, and causing a variation of the current flowing into the memory cell depending on the position of the cell. Compensation for this effect can be effected by controlling the gate voltage of the cell's selector transistor in such a way that its value is proportional to the source resistance.

So far, only a planar MOS transistor has been considered as a selector device, but this solution leads to a cell dimension that is quite large and is acceptable only in low-density memories for embedded applications. When considering larger memories, different selector devices can be used instead of planar MOS, resulting in less area occupation and very low leakage.

4 Ferroelectric Memory Architecture

Ferroelectric memories have borrowed many circuit techniques from DRAM's due to the similarities in their cells and DRAM's mature architecture. In particular, the same, very well-known array architectures of DRAM can be adopted also for FeRAM: they are the open bit-line architecture of Fig. 16a and the folded bit-line architecture of Fig. 16b. The first one, in which bit-lines are allocated on opposite sides of the sense amplifiers, allows a more compact array, but it is more sensitive to bit-line noise from word-lines, plate line and substrate and mismatches due to process variations with topology. Often the folded bit-line architecture is preferred which guarantees much better immunity from noise even if it is more space demanding. FeRAM cells are allocated in a folded structure as in Fig. 17. In this

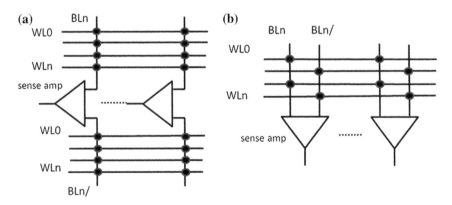

Fig. 16 Block diagram of a ferroelectric memory with **a** open-bitline architecture and **b** folded-bitline architecture

Fig. 17 PL tied with WL array scheme

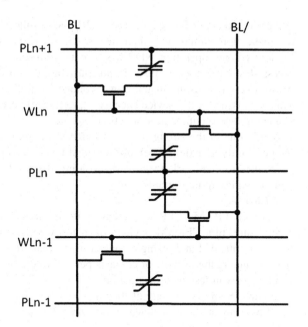

configuration, sharing the drain of two transistors it is convenient because their active area capacitance is reduced. Due to the fact that plate (PL) must be switched during cell operation it is convenient to implement multiple smaller PL lines instead of a single line common to all the array, so that the capacitive load to move during operation is reduced. Eventually a hierarchical PL decoding architecture can be used. A possibility is to link a single PL with a single word-line: in this case, referring to Fig. 17 enabling word-line WL_n would result in enabling also PL_n enabling all the cells sharing the same word-line.

It is interesting to consider the unselected cells that share the same plate line of the selected wordline, referring again to Fig. 17 for example the cells on the wordline WL_{n-1} when WL_n is selected: depending on their state those cells can be disturbed when PL_n is pulsed because the ferroelectric capacitor is subject to a voltage given by the capacitive divider formed with the parasitic capacitor of the storage node.

A dual decoding option is to run the Plate signal parallel to the BL. In this case, only the memory cell that is located at the intersection of a WL and a PL can be selected activating both the signals at the same time. This configuration is more flexible than the previous one, because it is possible to access only some of the cells in a row thus power consumption is reduced, but the disturb problem described above is still present, now affecting the cells in the same bit-line. As in a DRAM, it is possible to consider an architecture with a common PL biased at a constant voltage: the so-called-Non Driven Plate (NDP) architecture. Keeping the plate at a fixed bias, for example $V_{dd}/2$ if V_{dd} is the external supply voltage, eliminates the problem of delay connected to the high capacitance of plate line, but it allows a maximum voltage of $V_{dd}/2$ instead of V_{dd} across the ferroelectric capacitor. This can make more difficult to switch the ferroelectric capacitor and results in a problem for low-voltage operation NDP has also the drawback that when the cell is not accessed storage node is floated and must be kept to $V_{dd}/2$ to avoid it may fall below PL bias causing an unwanted voltage is applied to ferroelectric capacitor. This is an issue for the device because $V_{dd}/2$ must be periodically restored in all the storage nodes to compensate junction leakage.

While one of the strong points of FeRAM in comparison to DRAM should be the possibility to avoid refresh during stand-by, we can go further and consider the possibility to clamp PL to ground, avoiding the need for refresh, this requires a more complex timing architecture.

CFRAM

A chain FRAM (CFRAM) architecture is similar to a NAND-flash memory architecture. In a CFRAM, a chain of memory cells shares a single contact to the bit-line at one end and a single contact to the plate-line at the other. As in NAND flash, reducing the number of contacts per cell to the bit-line and the single contact to PL results in less area occupation.

Figure 18 shows the circuit diagram of a cell. In a stand-by operation, all the word-lines are at "high" voltage to guarantee 0 V across all the FE capacitors in the block.

In the active operation, a cell is accessed by grounding its corresponding word-line and raising the Block-Select (BLS) signal with all the other word-lines high, enabling the bit-line voltage and the plate voltage to propagate to the selected cell.

The first advantage of CFRAM is a reduced area with respect to mainstream 1T-1C approach. A second advantage is a reduced bit-line capacitance because only one cell of the string needs to be contacted to the bit-line (this is the main contribution to the total bit-line capacitance), which allows to increase the number of cells per bit-line improving array efficiency.

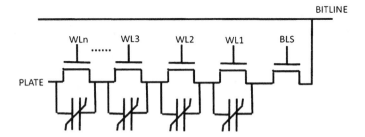

Fig. 18 Schematic diagram of a CFRAM architecture

Looking at Fig. 18 we can identify an issue of CFRAM related to the transistors in parallel with Ferroelectric capacitors: during operation they must short-circuit all the capacitors but the one of the selected cell. As they are not ideal components their drain-source resistance is not zero when they are driven ON so that a disturb voltage can appear across the unselected cells.

5   Cross-Point Array

From the previous considerations about 3D approach comes the idea of an array in which the memory cells are at the cross point of mutually perpendicular selection lines, namely the word-lines and bit-lines. This configuration realizes the maximum array density, and it is suitable to pack many memory layers one over the other for a 3D-memory configuration. It may lead to the minimum possible cell area of $4\,F^2$, considering a minimum pitch of 2F in X-direction and a minimum pitch of 2F in Y-direction, but, considering the possibility to stack multiple layers, the effective cell's size is $4\,F^2/n$ (where n is the number of stacked layers). A diagram showing in principle this architecture is presented in Fig. 19.

In addition, this configuration offers the possibility to locate the driving circuitry below the array (see Fig. 20), and so increasing considerably the chip's efficiency.

Placing decoders and sensing below the memory layers puts a constraint on the size of the tiles in which the memory arrays are partitioned: the required area for row decoders and column decoders is proportional to $N_{row} + N_{col}$ and increases linearly with the number of columns and rows.

On the other hand, the area size of a square tile is proportional to $N^2 A_{cell}$, if A_{cell} is the area of a single cell so, to allocate the decoders completely under the memory layer, the following relationship must hold:

$$NA_{wdec} + NA_{cdec} = N^2 A_{cell} \tag{12}$$

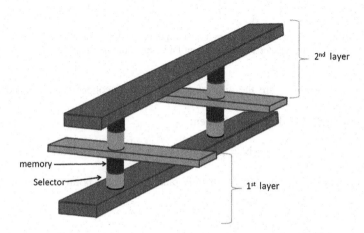

Fig. 19 Schematic of 2 memory layers in a cross-point array of a 3D memory

Fig. 20 Comparison of the allocation of memory array in a planar device and in a 3D device. In the last case, the array is shared in a number of "stacked" layers saving silicon space. Part of peripheral circuitry is placed beneath the memory layers

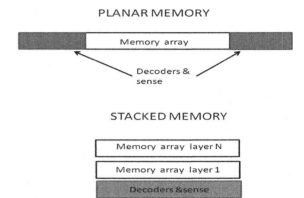

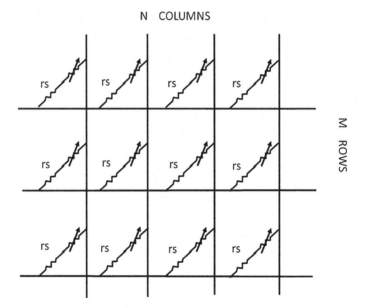

Fig. 21 Circuital diagram of a cross-point array (without considering parasitic elements). R_s is the resistance associated to selectors, and it is a non-linear function of voltage across it

where A_{wdec} and A_{cdec} are the area of a single row decoder and column decoder circuit respectively.

The minimum tile dimension is then:

$$N = (A_{wdec} + A_{cdec})/A_{cell} \tag{13}$$

The circuital representation of a single plane of a cross point array is shown in Fig. 21.

In the schematization shown in this figure, r_s stands for the resistance of the selector device and the memory element is represented by variable resistance, while the parasitic resistances of rows and columns are not considered.

This architecture is similar to the one discussed above for PCM, but here generic device with suitable I/V characteristic is considered as a selector. Under some assumptions for the selector characteristic, it is possible to adopt an array bias configuration that minimizes the total leakage during write, as shown in Fig. 22. In this figure, one assumes selection of the lower-right cell (in red in the schematic), and the corresponding bit-line is biased to write voltage V_w, while the corresponding row voltage is tied to ground (gnd) and all the other rows and columns are biased at $V_w/2$. It is easy to see that all the cells in the array can be classified with respect to their bias voltage V_{bias} into four classes: $V_{bias} = V_w$: the only cell in this condition is the selected one to which is applied the full write voltage

$V_{bias} = V_w - V_w/2 = V_w/2$: these are the cells located on the same bit-line as the selected one
$V_{bias} = V_w/2$: these are the cells located on the same word-line as the selected one
$V_{bias} = V_w/2 - V_w/2 = 0$: all the other cells in the array

Then the "V/2" bias scheme can eliminate the parasitic current through unselected cells during write except $N + M - 1$ cells which are biased at half the voltage required to write.

The sum of all these currents can be an important issue as we already know because it must not exceed the budget assigned to the array in the total memory stand-by current (I_{stb}) specification. Usually for large memories the greater part of

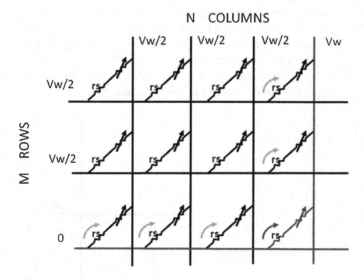

Fig. 22 V/2-biasing scheme

this current is concentrated in the array. The requirement described above can be expressed by the following relationship:

$$(N + M - 1)I_{off} < \beta I_{stb} \tag{14}$$

where $\beta < 1$ is the fraction of total I_{stb} allowed for the array. This condition can be extremely difficult to satisfy for a large array: for example, let's consider a 1 Gbit memory with $N + M = 1.5e5$, a maximum standby current of 150 µA for the chip at 80 °C operation temperature and $\beta = 0.7$. This means that about 100 µA can be allowed as the total stand-by current of the array, leading to I_{off} less than 660 pA/cell.

As we have already said a way to avoid or minimize this problem is to partition the array into smaller sub-arrays and activate only some of them at a time, but, as this technique is expensive in terms of silicon area and product performance, it is also required that the maximum effort is made to reduce selector sub-threshold current as much as possible, thus enabling minimizing the number of tiles into which the whole array is divided. The "V/2" scheme represented in Fig. 22 is not the only one that can be adopted, and other choices can be made to cope with the selector characteristic. It is clear from the discussion that a critical component in cross-point memory architecture is the selector device [4]: we can consider the requirements of an ideal device with reference to its I/V characteristic shown in Fig. 23. First of all, it should show a very high I_{on}/I_{off} ratio, on the order of 10e7 to efficiently reduce the leakage current of non-selected cells. Actually this ratio is linked to the I_{on} capability and I_{off} max that can be tolerated, which, in turn, depends on the dimension of the tiles into which the whole array has been divided.

I_{off} depends on the bias voltage of the selector, and various current specifications can be defined depending on the bias scheme used in the array to minimize the overall leakage contribution. If we consider, for example, the "V/2" scheme, we can request very low leakage near 0 V, because it is the condition of the majority of the cells in the array and allows a much higher current at V/2, where are positioned the "few" half-selected cells. To exemplify these ideas, we could choose 10 pA for the first case and 100 pA for the second one.

A high current density above a threshold voltage ($I_{on} > 10$ MA/cm^2) is also required to supply the current needed during programming.

Fig. 23 Typical non-linear characteristic of a selector device

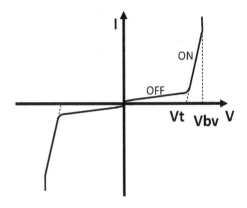

The threshold voltage to turn on the selector is a critical parameter: it cannot be too low to avoid unwanted turn-on of some cells, but it cannot be too high because this would lead to un-manageable high voltage operating on the array.

To allow bipolar operation, like in ReRAM and MRAM, a symmetrical characteristic is mandatory, even if some small asymmetry can be tolerated, this will impact on the worst case condition of array biasing.

Finally, as the selector is intended to be stacked with the memory element in a 3D array, a back-end compatible process (<400 °C) is required for its fabrication.

A large number of experimental devices have been considered to be used as a selector in cross-point/3D memory arrays: P-N diodes that we have seen previously are used in some memory architectures, such as PCM, with mono-directional operation principle, polysilicon diodes have been proposed for 3D stackability but the need for processing at temperatures <400 °C could be an issue. The next class comprises experimental devices designed to attempt having all the characteristics requested for an ideal selector, first of all, a strong non-linear I/V characteristic.

- Metal-oxide Shottky barrier devices: are based on the principle of creating Schottky contact by metal/semiconductor interface engineering, using ultrathin silicon layers. They are relatively simple to integrate, but the current driving capability should be improved.
- Oxide tunnel barrier: one attractive approach for achieving strongly nonlinear I-V characteristics is to use a thin oxide or nitride layer as the tunneling barrier. Tunnel barriers can be formed by various high-k materials (HfO_2, SiO_2, ZrO_2, and TiO_2) and the field sensitivity can be enhanced by tunnel-barrier engineering of multilayer oxides. According to how the dielectrics are stacked, such engineered tunnel barriers are classified as crested and variable-oxide-thickness (VARIOT) type.
- Mixed-ionic-electronic-conduction(MIEC): Copper-containing MIEC materials have become an interesting choice as 3D-ready access devices for NVM. MIEC materials can be processed at temperatures below 400 °C, making them back-end-of-line (BEOL) friendly. MIEC-based access devices offer large ON/OFF ratios ($>10^7$) and low leakage (<10 pA around 0 V and <100 pA @ 0.4 V). In pulse-mode, the devices can carry the high-current densities needed for PCM and fully bipolar operation needed for high-performance RRAM, and their endurance is very high ($\gg$1e10). As a drawback, endurance is strongly dependent on the current (programming current) and voltage across the memory element should be low (1–2 V).
- Finally, devices with a threshold switch are very interesting to consider (see Fig. 24). A threshold switch makes for an ideal access device, since, from an initial highly resistive state, the device switches to a highly conducting state as soon as a threshold voltage or current is applied. Various types of threshold-switch-based devices such as Ovonic Threshold Switch (OTS), which makes use of the properties of chalcogenide itself to obtain a switching characteristic,

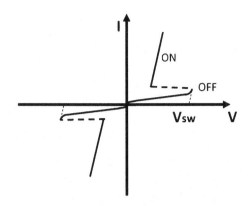

Fig. 24 Typical characteristic of a switching selector device

Table 1 Pros and cons of various types of selector devices

Selector	Current density (on)	Selectivity	Bidirectional	3D
Vertical bjt	Fair	Yes	Yes	No
P-N diode	Good	Yes	No	Fair
Chalcogenide threshold switch	Good	Fair	Yes	Yes
MSM (schottky barrier)	To be improved	Yes	Fair	Yes
MIEC	Good	Yes	Yes	Yes
Oxide tunnel barrier	Good	Yes	Yes	Yes

metal–insulator-transition (MIT), and threshold-vacuum switches have been proposed for application as access devices in cross-point arrays.

An interesting option using OTS has been proposed for PCM technology [5]. The fabrication process of OTS makes it feasible to stack PCM + OTS, to make a cell which can reach the minimum dimension $2\,F \times 2\,F$. The memory array can then be a cross-point array configuration as previously shown (see Fig. 19) with a $4\,F^2$ cells in every node. To understand the electrical characteristic of the whole cell, we can start from the characteristics of single components and make the sum of them, utilizing the rules of series connected bipoles. Both of them have a switch-back (negative resistance) in their characteristic, and this is reflected in the final I/V curve that will show two switching thresholds, one related to OTS and the other related to GST material. To retrieve data stored in the cell, it will be necessary to apply enough voltage to switch OTS on, avoiding an unwanted switch of the memory element. This requires to carefully design the threshold switch of the two elements and results in a more complex design of the reading circuitry. Investigation into optimum selector devices suitable for use in large memory arrays is on-going, and it will be the critical factor for feasibility of 3D memory using emerging technologies; the following table summarizes the pros and cons of some types of selector devices (Table 1).

6 Write Circuits

6.1 Introduction

A very interesting feature of emerging memories compared with mainstream flash technology is the possibility to modify a single bit, instead of being forced to operate on a whole sector of the array, and a high write speed. This opens the possibility to simplify garbage collection management and application software to make it faster, and, also, to achieve a read/write speed at least competitive with the slowest DRAMs. The other hope was that the complex algorithm needed, in particular, in NAND flash to achieve a compact distribution of written cells could be avoided or at least greatly simplified in the new technologies obtaining an obvious advantage in the write speed but also producing a big simplification of product design due to the elimination of write state machine necessary to manage the write algorithm. To this extent, there are remarkable differences among the various approaches in the emerging memory area, some of them like PCM, for example, in which the write mechanism has a certain degree of graduality still take advantage of a detailed writing algorithm to achieve a satisfactory cells distribution in big arrays. In other cases, the writing process is more similar to a switch phenomenon that is difficult to control with an algorithm.

6.2 Writing PCM Memories

As we have seen, the physical mechanism used to change the phase of chalcogenide material (GST) between amorphous to crystalline and vice versa needs to heat the material using Joule effect. If a reset is being done, an electrical pulse causing enough current flow through heater and chalcogenide can rapidly reach the melting point of a small portion of material above the heater (Fig. 25), causing a transition of the chalcogenide to an amorphous state. Then, the amorphous condition of GST material must be "quenched" to remain unchanged for a long time against the natural trend to return to crystalline. This, as we know, can be obtained with a

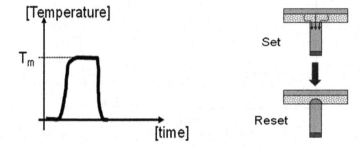

Fig. 25 Reset operation in a PCM cell

sudden removal of the current through the material, and, to this extent, the trailing edge of the RESET pulse must be extremely fast, on the order of 2 ns.

The constraint on the trailing edge of the reset pulse caused a problem for the designer because write voltage is applied to the edge of array bit-lines and then is subject to a time constant determined by parasitic distributed R and C. Both these values are proportional to length, so the time constant increases with the square of bit-line length. Then instead of the usual pulse control on the bit-line, it could be more effective to control the word-line to switch-off the current, since the capacitance associated with the selector is less than the bit-line capacitance.

Even if, on a single cell, a pulse width of 10–12 ns proved to be enough for a safe reset, the need to have a margin for statistical distributions in a large array increases this number to a level of approximately 50 ns.

When we want to switch the cell to a condition of low resistance (set), it is necessary to change the material phase from amorphous to crystalline. As we know, this is done by heating the material to a temperature which is lower than the one reached during reset, followed by a relatively slow freezing to provide time to initiate the growth of crystal nucleation points in the material. From the point of view of the designer, this means to apply a lower current pulse with a longer width (see Fig. 26).

The main challenge for the SET pulse is to match the right crystallization temperature that is dependent on the cell characteristics and its parameters spread. To maximize the efficiency of the operation, it can be convenient to adopt a slow trailing edge at the end of the pulse. This guarantees that, even if the right crystallization temperature is lower than expected, the pulse will intercept it and stay around long enough to ensure that the crystallization process takes over: in other words, it ensures a better tracking of the cell's variations due to technology. A linear slow-down is probably the simplest solution, and the width of the triangular section of the pulse is dominating the whole pulse width (or it could be a trapezoidal [6] pulse like depicted in Fig. 27).

To maximize yield, a program-and-verify algorithm must be adopted in SET/RESET operation. Basically this means that after every RESET or SET pulse,

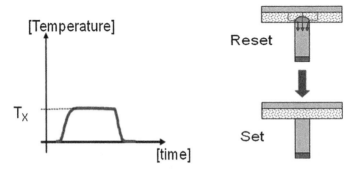

Fig. 26 Set operation in a PCM cell

Fig. 27 Possible shape of a
SET pulse

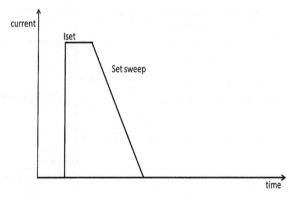

a read is performed to check if the selected cell is modified or not. Of course, this
will slow the SET/RESET operation.

During SET/RESET operation, the relationship between the temperature reached
into GST and the power delivered to it is:

$$\Delta T = P \times R_{th} \qquad (15)$$

where ΔT is the temperature increase, P is the power applied and R_{th} is the thermal
resistance, depending on the characteristics of the material. In spite of the fact that a
certain power P is delivered from the write pulse, this power is distributed in a
different way among the cells of the array: depending on their position in the array
cells are subject to different parasitic resistances, then the power dissipated in
parasitic elements is different. Moreover memory element interface area shows
variations due to fabrication process, the composition of GST may present local
variations that change slightly its properties so that different cells may require
different power to reach the same temperature increase. The target of design is to
ensure to every cell enough power to guarantee a safe write minimizing the effects
of power distribution along the array. If the write circuit is designed as a voltage
generator (i.e., to force a voltage), it will result in a large variation of applied power
with position, parasitic resistance variation, and a significant sensitivity to the
interface area of the memory element.

If the write circuit is designed as a current generator (i.e., forcing a current), the
impact of parasitic resistance variation is cancelled, but the impact of the variation
of the interface area of the memory element is larger and a higher voltage is
required for the write driver. Then the choice of the right biasing mode is dependent
on what is the worst case of parameter distribution.

A source follower is often used to apply write voltage to the array, enabling
stable voltage regulation not affected by the current sinked by the cell. In this way,

Fig. 28 In the current forcing method, SET and RESET current are forced in PCM cell

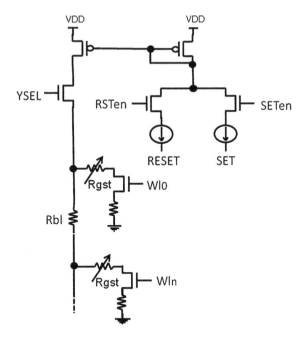

Fig. 29 GST resistance as a function of applied current

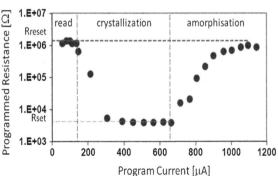

it is easy to vary the applied voltage with the different operation modes (set, reset, read) [7]. A principle schematic of this writing scheme has been discussed previously (see Fig. 12).

Using the current force method, a current mirror circuit can be used to fix the current flowing into the cell as in Fig. 28. Different current references enable management of SET and RESET operations [8].

Since the beginning of PCM development, the possibility to program the cell to various resistance levels, making possible multilevel operation, has been investigated due to its importance to compete with NAND products which can implement this feature. The possibility to achieve a continuum of resistance values can be envisioned by Fig. 29.

In this graph, you can recognize that the slope of the resistance change passing from crystalline to amorphous state is smooth enough to allow one to tune the programming current to obtain an intermediate resistance value. "Analog" programming of PCM cells can be obtained by controlling the height and duration of the programming pulse for a rectangular current pulse. Also triangular-like pulses, like those used in the set operation in which the slope is variable, can be used. In general, in any case, to tune the value of resistance with enough precision, a dedicated program and verify algorithm have to be used. But before talking about this, it is necessary to review some phenomena affecting PCM cells that are relevant for single-level programming, but may become a big barrier in multilevel programming. First of all, the fact that an intrinsic variability is associated with each write pulse means that the value of resistance after the pulse is not exactly repeatable. The second and most important point is a phenomena called short-term drift that shifts the resistance of a just reset cell toward higher values [9]. This phenomenon is particularly dangerous in multilevel programming because intermediate levels are more affected, and, in the end, it limits the possibility to increase the number of storage levels.

Short-term drift is a slow but steady increase of the resistance of amorphous materials, and the resistance change can be described by a power law:

$$R(t) = R_0(t/t_0)^\alpha \qquad (16)$$

where $R(t)$ is the resistance at time t, R_0 is the resistance at time 0, and α is a power coefficient that has been found to be on the order of 0.05–0.1 for amorphous GST. Short-term drift has been explained in terms of structural relaxation in the amorphous material that modifies Poole-Frenkel conduction through GST, and also in terms of conduction through electrically active defects inside the material.

There is also long-term drift visible on reset cells that is caused by the fact that as we know the cell tends to return to its stable state which is the crystalline one. This can be similar to the retention charge loss of the familiar flash technology, of course with a different cause. Finally, all the variables associated with the process and geometry in the array contribute to create a working spectrum difficult to control with multilevel programming and limiting the number of levels that can be accommodated in a cell: 4 levels (2 bits/cell) and 16 levels (4 bits/cell) have been demonstrated in prototypes, but not in full production so far.

Conceiving an algorithm for multilevel programming requires deciding if one start from a completely set device or from a reset one, and this means a different model of the resistance-change process.

In the case of starting from a set device, the idea is to grow the volume of amorphized material using an increasing level of reset current. In a simplified view this growth is supposed to proceed by increasing the radius of the spherical amorphized volume, as in Fig. 30.

Following this model, the algorithm should be a full set pulse followed by increasing-width reset pulses, followed by a verify after each of them.

Fig. 30 Intermediate resistance levels obtained by an expanding amorphous region

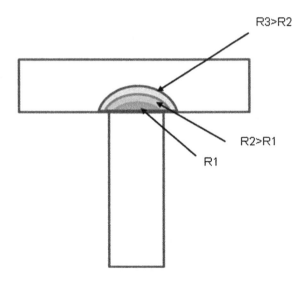

Fig. 31 Formation of a conduction filament in amorphous GST during set pulse. Lower resistance values correspond to larger filament size

The other possibility is to start from a fully reset cell. In this case, small set pulses create crystallization nuclei that open a crystallization path inside the amorphous zone. This path is then enlarged by the following set pulses as in Fig. 31.

6.3 Writing ReRAM (Bipolar) and STT-MRAM

ReRAMs and STT-MRAMs writing requires the possibility to reverse the polarity of the voltage applied to the selected cell depending on the operation to be executed. Both these technologies are based on a memory-material layer sandwiched between a top and a bottom electrode and basic write circuit is in principle simple because a pulse of appropriate height, duration and polarity should be applied across the electrodes. Unlike PCM, which has some constraint on leading edge and trailing edge of the writing pulse due to phase change phenomena, simple pulses can be generated. In case the architecture with fixed plate can be used, it is enough to drive the bit-line to V_{write} or Gnd, while the plate will stay at a fixed bias voltage

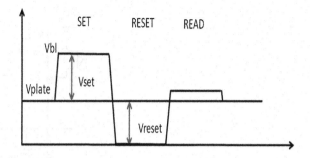

Fig. 32 Changing bit-line voltage with respect to a fixed plate, it is possible to reverse the bias of memory cell. Although this appears to be an easy solution, stress on selector device is worsened

and the voltage applied to the cell in SET, RESET and READ conditions will be as in Fig. 32.

In case an array architecture with alternate bit-lines and source is chosen, it is necessary to provide a circuit-driving bit-lines and related source lines alternatively at high voltage for set and reset operations. In other words, bi-directional driver circuits driven by the data to be written are required (see Fig. 33). The figure depicts an example of bi-directional driver for a STT-RAM cell. Data signal D_1 and D_0/activate the write buffers, then if a zero is going to be written, voltage is applied to SL_n through M_3 (D_0 to gnd), turning on sink device of BL_n (M_2) which sinks the current from BL to ground (see Fig. 33a). If a 1 is going to be written, voltage is applied to BL_n through M_4 (D_1 to gnd) turning on sink device of SL_n (M_1), which sinks the write current from SL to ground (see Fig. 33b). A Write Enable signal, not shown in the figure, should enable buffers operation [10].

An important issue rising during programming is the possibility to stress or even break the memory stack by applying too high a voltage. It is important to note that the cell in the array that requires the higher voltage to write due to topological reasons (resistive drop on the bit-line and/or bit-line), or to the spread in process parameters, will determine the voltage for all the other cells. To avoid or mitigate reliability problems that could rise from this situation, a staircase voltage can be adopted so that, at each step, the cell is verified and the staircase stops as soon as the cell passes the verify condition. In this way, each cell receives only the minimum voltage required to write. The drawback is that the write latency is much higher due to the need for verification.

In STT-RAM, it is necessary to have a writing current I_w higher than a critical current I_c flowing into the MTJ because the probability of switching of the free layer is a sharp function of critical current. The optimum condition is to maintain I_w very near to critical current value because higher values bring a higher risk of junction breakdown and the need for a larger selector transistor that increases the cell area. As the choice of writing current affects the switching probability of free layer of the MTJ, the required write error rate (WER) is an important design specification that must be taken into account, and, depending on the required WER, it is possible to decrease I_w.

Like in PCM, it is useful to analyze what is more convenient between forcing the write current or forcing a voltage high enough to make the writing current flow into

Fig. 33 Write driver for a
STT-RAM cell **a** write 0,
b write 1

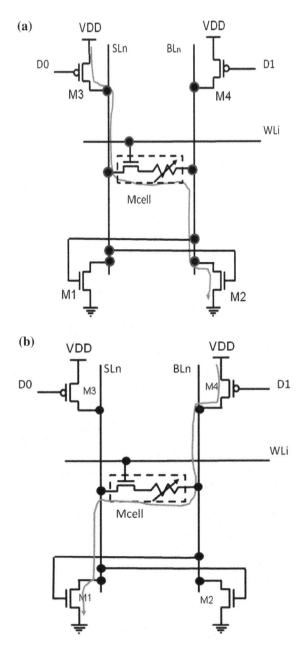

Fig. 33 Write driver for a STT-RAM cell **a** write 0, **b** write 1

the cell. In case of voltage forcing, even assuming a perfect voltage generator, it must be remembered that it cannot be applied directly across the MTJ, but only at the top of the selected bit-line. Also, of course, the parasitic resistance of the bit-line itself, the R_{on} of the cell selector. and the source resistance R_s to ground act as a series resistance with the voltage generator.

The voltage applied to MTJ is then given by:

$$V_{wmtj} = V_w R_{mtj}/(R_{on} + R_{bl} + R_s + R_{mtj})   \qquad (17)$$

where R_{mtj} is the resistance associated with MTJ junction and R_{bl} is the parasitic resistance of the bit-line up to the contact of the selected cell. Considering R_{on} dominating over the other contributions the relation above can be approximated as:

$$V_{wmtj} = V_w R_{mtj}/(R_{on} + R_{mtj})   \qquad (18)$$

From the above relationship we can see a potential issue if R_{on} is comparable to the other contributions during a MTJ transition from parallel state (Rmtj at low resistance = R_p) to anti-parallel (Rmtj at high resistance = R_{ap}), because an over-voltage can be produced across the MTJ.

Overvoltage is minimized if R_{on} is made as small as possible, but this requires a larger selector area, which will increase the cell dimension. If a voltage force technique is chosen, it is necessary to provide a current limitation to avoid that, due to the process or geometrical spread, too high a current can flow in some of the cells, also the sensitivity to bit-line and source-parasitic resistance must be considered, and some kind of compensation should be provided.

The other possibility is to drive the write circuit forcing a suitable current into the cell. Overvoltage generated with this technique during the switching $R_p \rightarrow R_{ap}$, is not dependent from R_{on} but only from R_p, TMR, and I_{cell}. To avoid the risk of reaching MTJ breakdown voltage, a voltage limitation should be provided. Another point is that in this case there is no sensitivity to the bit-line parasitics.

Also in the ReRAM (for example CB-RAM), a great care must be taken in controlling the current flowing, in particular, when going toward the low-resistance state (set current).

Some current must then be sustained for a defined time for the cell to become reliably set. This current must be carefully controlled, because if it is too low the cell will not be well set and may present a data retention issue; on the other hand, if it is too high, the cell may experience an "over-set". In this condition, it is very difficult or impossible to recover the cell to the reset condition; then the I_{set} provided to the cell must be placed between two levels determined by previously described constraints that we can call $I_{set_{min}}$ and $I_{set_{max}}$. Unfortunately, a transient phenomenon that occurs when the cell switches from high resistance to low resistance makes it difficult to maintain the correct value of current at all times through the cell. To explain the problem, let's refer to the model of a bit-line in a memory array to which a memory cell is connected (see Fig. 34). The bit-line is characterized by a parasitic resistance and a parasitic capacitance distributed through the length of the line but concentrated in two components in the model; the capacitance C_{bl} is charged at a fraction of the set voltage at the top of the bit-line as determined by the resistive divider made of the portion of bit-line resistance up to the memory cell considered and the resistance of the memory material. At the

Fig. 34 Circuital
schematization of a ReRAM
or STTMram cell connected
to a bit-line

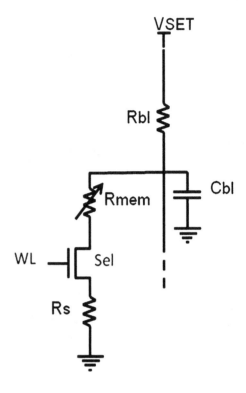

beginning of the set operation, this resistance is usually high (R_{hi} in the range of many hundreds of kΩ) and the capacitance is charged at the value:

$$V_{cap} = V_{set}R_{hi}/(R_{hi} + R_{bl}) \tag{19}$$

But as soon as the filament is formed, the resistance of the memory element suddenly drops to R_{lo}, determining a change in the voltage divider ratio, due to the high R_{hi}/R_{lo} ratio present in CB-RAM technology, which can be on the order of 1000. This change would require a sudden and consistent reduction of the voltage across the memory cell, but, due to the parasitic capacitance, the bit-line cannot react immediately to change the voltage, and, as a result, a current overshoot is applied to the memory material until the bit-line capacitance has been discharged. Recovery time depends on the time constant $R_{bl}C_{bl}$ of the bit-line and parasitic elements associated with the memory cell.

This extra-current must be avoided to prevent damage to the cell and reduction of reliability, and, even if the source resistance of the selector can contribute with a negative feedback regulation effect, in general it is necessary a regulation circuit to compensate for the dangerous current overshoot [11].

A possibility to achieve this target is to use the cell selector which is usually an N-MOS device as a current source, operating it in saturation region: regulation of gate voltage, since the MOS is close to the cell, is an effective ballast. It will limit

Fig. 35 Set-current
regulation using a current
generator on top of a selected
bit-line. Selector is operated
in the linear region to
minimize voltage drop

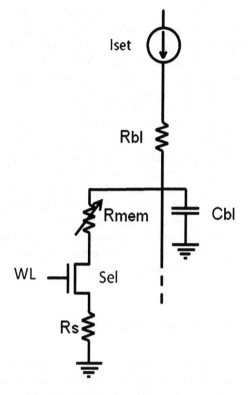

the transient current spike from the bit-line capacitance, but the selector MOS is usually very small to make the cell's area as small as possible, so its dimensions are not very well controlled and it cannot perform a perfect regulation of the current.

A widely used technique to control the current through the cell during set operation is to force a compliance current in the bit-line using a current generator like in Fig. 35.

Since the current source can be implemented by large devices, they are much better matched and controlled and this will result in a much tighter DC current distribution. On the other hand, since it is outside the array, the current source is not very effective to prevent the current spike provided by the bit-line capacitance.

7 Redundancy

7.1 Introduction

One of the most important issues of semiconductor devices fabrication is to max-imize fabrication yield for the obvious reason that the higher is the yield, the lower is the fabrication cost of the product. This is even more important in regard to memories because most of the competition in the memory market is focused on price; memories are mainly a standardized product without great differences between various vendors. Premium price is paid at the time of introduction of a novel product, but this effect rapidly decreases with time and it is quite small 1–2 years after its introduction.

Considering a memory chip, it is possible that, among millions or even billions of cells, some of them is found defective at the end of process fabrication, since fabrication yield is dominated by silicon defectivity which is related to process complexity (number of operations, number of layers, number of masks), which can be reduced by improving the cleanness and optimizing the production process and the technology. Other yield detractors can be detected in some weakness at the design level, which can be eliminated with a design refine.

To avoid discarding parts containing a small number of defective cells, some spare rows and/or columns are usually included in a memory chip. These elements can be activated during the electrical wafer sort and used to replace one or more defective elements in the array.

7.2 Redundancy Schematic

In Fig. 36, a conceptual schematic of a redundancy circuit for a generic memory is shown. In this case, the circuit replaces a row of the array with a defective cell (red circle in the figure) with a spare row of working cells (bottom row). To do this, it is necessary that, when the address of the row containing the failed cell (RADD2) is present in the row address bus (RADD), a signal DIS is generated that disables the normal address decoders and activate the spare row. As shown in the figure, this is

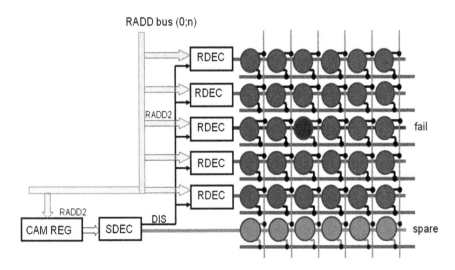

Fig. 36 Schematic of row redundancy for a generic memory

achieved by programming the spare row decoder (SDEC) with the address of the failed row as identified during the electrical wafer sort: in this way, when the address ADD2 will be present on the address bus, it activates the SDEC and the disable signal for the normal decoders.

As the address of failed rows cannot be known until the testing program has detected it, the fail address must be written directly into a non-volatile register (CAM in the figure) by the testing equipment. This register is one of the most critical items of the redundancy design because a failure in reading at this level will cause a malfunction in the whole memory. Laser fuses or electrical fuses have been commonly used as programmable elements, but, if the memory is a non-volatile memory, some attempts also have been done to use non-volatile cells as programmable elements for redundancy register in spite of the higher risk of failure, because fuses (in particular, laser fuses) are expensive in terms of area.

Looking at the example of Fig. 36 is immediately clear that redundancy is expensive in terms of area: to repair a single fail cell all the cells of a row have been replaced; if the array is partitioned in tiles like happens in large memories it is possible to substitute rows in a single tile reducing the area overhead.

Usually, spare columns are also inserted into the array to increase the capability to correct errors but also to optimize the use of spare elements: For example, if many bits are failed in a column, it is easy to repair the chip using a single column while it would be impossible or very expensive to do it having spare rows only. The concept of spare-column decoding is similar to what has been explained for rows: in general a block of spare column is added in parallel to the array columns but here, to avoid to waste time during reading, a separate sense amplifier is used, and a selection signal makes the choice of the data to send to the output buffer. It is possible to replace a block of columns or a single column, and it is not necessary to make the substitution on all the output bits, but it is possible to do it only for the output bit that failed. A column redundancy schematic diagram is shown in Fig. 37.

The number of spare elements to be utilized in a chip is a compromise between the expected yield improvement and the area overhead. Increasing the number of spare resources above a certain point doesn't increase the electrical wafer sort yield anymore, because the larger silicon area increases also the failure probability. The curve of the probe yield as a function of the number of repair resources available can appear as in the drawing of Fig. 38. Generally the area dedicated to repair elements is <10% of total chip area.

As a last consideration, it must be noted that the best use of spare elements (rows or columns), in relation to the kind of failures found on silicon (redundancy strategy), is an important optimization consideration to be done in the testing software to maximize silicon yield. To understand this, we can take a trivial example of bit pairs' failures: if they are aligned on a column, only one redundancy resource (a column) is enough to repair the part, but, if spare rows are used, we need two of them; if they are aligned along a row, it will be just the opposite.

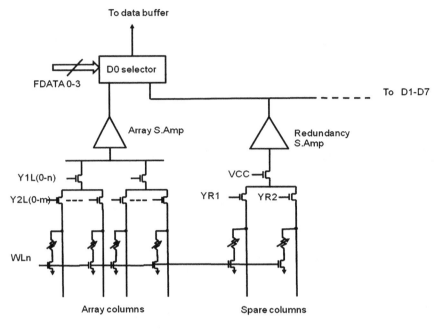

Fig. 37 Simplified column-redundancy architecture

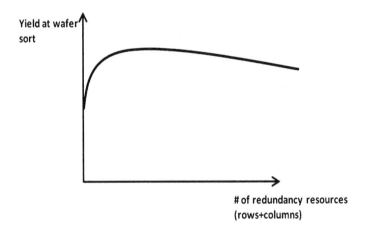

Fig. 38 Drawing of the relation between Yield at wafer sort and number of spare elements implemented in the chip

References

1. F.Bedeschi et Al. "A multilevel cell bipolar selected Phase Change memory" ISSCC2008, Digest of technical papers
2. J.S.Meena et Al. "Overview of emerging non-volatile memory technologies" Nanoscale feas. Letters, 2014,9:526
3. R.Fackenthal et Al. "A 16Gb ReRAM with 200MB/s write and 1GB/s read in 27nm technology" ISSCC2014 Digest of technical papers, 19.7
4. G.Burr et Al. "Access devices for 3D crosspoint memory" Journal of Vacuum Science and Technology B,Vol 28, issue 2, pp.223-262, March/April 2010.
5. D.Kau et Al. "A stackable cross point phase change memory" IEDM2009, Digest of technical papers;27.1.4
6. Y.Choi et Al. "A 20nm 1.8v 8Gb PRAM with 40MB/s program bandwidth" ISSCC2014 Digest of technical papers
7. F. Bedeschi et al."An 8Mb demonstrator for High-Density 1.8V Phase-Change Memories" Symposium on VLSI circuits,2004, Digest of technical papers; 26.1
8. H. Chung et Al. "A 58nm 1.8v 1Gb Pram with 64MB/s program bandwidth" ISSCC2011 Digest of technical papers.
9. G.Burr et Al. "Phase Change memory technology" Journal of Vacuum Science and Technology B Vol.28, issue 2,pp.223-262; March/April 2010
10. T.Kawahara et Al. "2Mb SPRAM(SPin-Transfer torque RAM) with Bit-by-bit Bi-Directional Current Write and Parallelizing Direction Current Read" ISSCC 2007, Digest of technical papers
11. S. Dietrich et Al. "A Nonvolatile 2 -Mbit CBRAM Memory Core Featuring Advanced read and Program Control" IEEE Journal of Solid State Circuits, Vol.42, No.4, April 2007; pp.839-844

Data Sensing in Emerging NVMs

Roberto Gastaldi

1 Introduction

Emerging memories currently in development in the electronics industry have in common the fact that the signal to be detected is a change of electrical resistance of the memory material, even though the way in which this resistance variation is obtained is different for each of the technologies considered. This is the case in PCM, Re-RAM, and MRAM, with the only exception being FeRAM which is very similar to DRAM from the point of view of the sensing concept. Then it is possible to associate a binary value for example "1" to low resistance and the other binary value "0" to high resistance, while "high" or "low" resistance is defined by a comparison with a reference resistance value. The sensing challenge is then to decide in a fast and reliable way if $R_{cell} > R_{ref}$ or $R_{cell} < R_{ref}$.

FeRAM instead can be sensed in a way similar to DRAM except that charge sharing is done with two different equivalent capacitances for "0" and "1". Unlike DRAM, a reference capacitance must be built midway these C_0 and C_1, and this poses some problems for designers.

Before describing the sensing circuits used in various emerging memories, let's recall the sensing principles of flash and DRAM to underline the differences between them and that of the cells of emerging techniques.

R. Gastaldi (✉)
Redcat Devices s.r.l, Milano, Italy
e-mail: r.gastaldi@redcatdevices.it

© Springer International Publishing AG 2017
R. Gastaldi and G. Campardo (eds.), *In Search of the Next Memory*,
DOI 10.1007/978-3-319-47724-4_6

2 Sensing Concept in Flash and DRAM Memory

In a flash cell, information is stored in a floating gate. The charge stored in the floating gate causes a change in the flash transistor threshold and then of the drain current [1, 2]. The bias condition of flash during reading is such that the transistor operates in a saturation region so that the drain current depends mainly on the threshold (at fixed gate voltage). In fact, the drain current of the cell can be described by this approximate relationship:

$$I_d = KW/L(V_g - V_{th})^2 \tag{1}$$

where K is the transconductance gain and W/L is the dimensional ratio of the cell. Sensing means to compare the cell's current with a reference current. Actually I/V conversion is usually performed using a column load made of a current mirror and a cascode amplifier as a converter stage. The cascode is needed to control drain voltage and provide low input resistance to the cell and voltage gain to the comparator to enhance speed.

With reference to Fig. 1, we see a concept schematic of a sense circuit for a NOR-flash device. The control gate of the memory cell is connected to the array word-line which is at a voltage VWL in the selected row, while the selectors of column decoding are included in the column load. The cascode devices M_1, M_2 are connected to a bias voltage through a feedback amplifier (not shown in the figure) that regulates the drain voltage of the cell, reducing input resistance and increasing speed at the expense of an increased power-consumption and area. The column

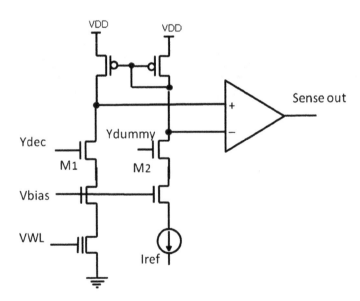

Fig. 1 Concept schematic of a flash-memory sense amplifier

decoding path is replicated in the reference tree to provide matched paths regarding resistance and capacitance for the array and reference legs. Voltage regulation is set to the higher drain voltage compatible with read disturbance, to increase read current and speed. The current mirror load improves amplification of small changes of the cell current due to high output impedance [3].

A reference leg is usually provided by a reference current based on a reference flash cell to be able to track average array cell behavior for process, temperature, and voltage variations. The reference cell has to be suitably programmed to generate a reference current midway between erased and written cell to have enough margins to detect both conditions, and this is usually done at the factory during probe testing. In the simplest scheme, a reference cell for each sense amplifier is provided, but this method is not practical when a high number of sense amplifiers is involved because the number of reference cells may grow too much, resulting in large area occupation and a programming operation that requires too much testing time. A method to share the same reference among N sense amplifiers is shown in Fig. 2. The reference level generated in a single location is mirrored two times and transferred to the reference leg of N sense amplifiers through an N-channel transistor. This method has the advantage that the reference level can be distributed to a high number of sense amplifiers without degradation, while the precision of the reference current can be well controlled and the capacitance C_{line}, made of the parasitic capacitance of metal connections, contributes to reduce disturbances on the voltage driving the reference transistors [4–6].

The sensing concept in DRAM is based on charge sharing between the cell capacitor and the bit-line capacitance [7, 8]. Cell capacitor is charged at $+V_c$ or $-V_c$, depending on the data stored in the cell; when the cell capacitor of the selected cell is connected to the bit-line through selector transistor, the bit-line voltage will change slightly due to charge sharing (Fig. 3).

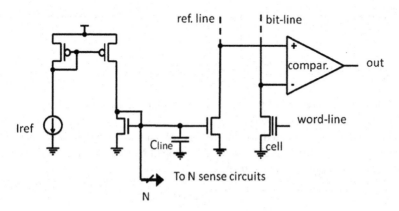

Fig. 2 Sharing the reference level among N sense amplifiers

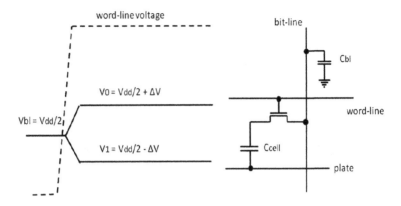

Fig. 3 Charge sharing in a DRAM cell

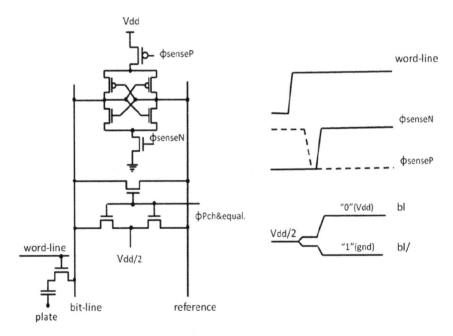

Fig. 4 Reading schematic of a DRAM-memory sensing scheme and the related timing

This change is compared to the voltage of an unselected bit-line used as a reference so that a differential voltage is developed between selected and reference bit-lines as shown in relationship (2).

$$V_{signal} = V_c(C_{cell}/(C_{cell} + C_{bl}))$$ (2)

The comparator used to detect this voltage difference is essentially a pair of cross-coupled inverters connected to Bl and Blref as shown in Fig. 4.

At the beginning of the reading cycle, the bit-line pair is equalized and connected to Vdd/2 through the precharge circuit, while the cross-coupled pair of inverters that is the sense amplifier circuit is disabled. As soon as the selected word-line connects the array cell to the bit-line through the cell pass transistor, a small signal (100–200 mV) starts to develop on the bit-lines pair; finally, when the sense amplifier is enabled by ΦrefP and ΦrefN, due to the high gain of this circuit, the bit-line is driven to Vdd or Gnd depending on the data stored, while reference bit-line is driven to the complementary value. This also provides a simple way to restore the data just accessed, because charge sharing destroys data written in the accessed cell. The concept of charge sharing is retained in ferroelectric memories (FeRAM) whose design can borrow much from DRAM design as we will see in the following section.

3   The Concept of Read Window

To guarantee a stable and reliable reading of the information stored in a memory cell, it is necessary that adequate separation exists between the two values of a parameter used to represent the logic data "0" or "1" (here we don't consider the case of multi-level storage, but the considerations we will make are easily applied also to that case) [9]. In a real case, we should talk about two distributions of values and it is then necessary to define two limits at the edge of these distributions to decide the logic state of a bit. For example, if we consider the current flowing into a memory cell, we can decide that logic "1" is represented by $I_{cell} > I_H$ and logic "0" is represented by $I_{cell} < I_L$, where I_H and I_L are two thresholds chosen after a careful characterization of the capability of the particular technology chosen for memory material. During the programming operation, a periodical verification is made of cell current with respect I_H or I_L depending on the data to be written, until one of those conditions is met, at which point the programming operation is stopped. Ideally, assuming adequate process capability, the window available for reading is exactly the difference $I_H - I_L$, but in a real case there are some factors that must be considered in calculating the signal window available for reading.

The first point is that the precision we can achieve in setting the limits I_H and I_L is limited, due to the variability of the circuitry to generate these reference currents with process, voltage, and temperature conditions and the parasitic elements introduced by the layout. The second point is that the circuit that makes the comparison between I_{cell} and I_H or I_L needs a minimum difference to output reliable result, so that, in the worst case, a quantity ΔI_L must be added to I_L and a quantity ΔI_H must be subtracted to I_H to obtain the real available read window.

Due to all these phenomena, the actual verify limits for set and reset operations become (see Fig. 5).

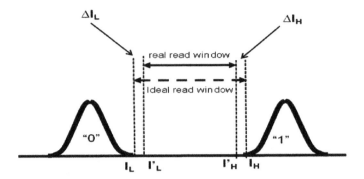

Fig. 5 Array distributions after applying an array program and verify algorithm

$$I'_H = I_H - \Delta I_H$$
$$I'_L = I_L + \Delta I_L \tag{3}$$

and the actual read window of the cell is:

$$\text{Read Window} = I'_H - I'_L \tag{4}$$

The reference level for the read operation should be set at the midpoint of the read window:

$$I_{ref} = \left(I'_H + I'_L\right)/2 \tag{5}$$

However, we have to consider again the limited sensitivity of the circuit used to make the comparison between the value of the current of a cell and the current reference as done before. The uncertainty zone is dependent on the design of the comparator and its sensitivity to process, temperature, and voltage variations; in addition, the value of the reference current can vary arising from the spread of electrical parameters of the reference generation circuit, so then also the reference current will have a distribution of values. To design in the worst case condition, the average value of $I_{ref} \pm 3\sigma$ must be considered. If we call $+/-\Delta_{sense}$ the sum of 3σ variation of I_{ref} and the guard band due to the sensitivity of the comparator, to have a reliable reading, the following relationships must be verified (see Fig. 6):

$$\left(I_{ref} - \Delta_{sense}\right) - I'_L > 0$$
$$I'_H - \left(I_{ref} + \Delta_{sense}\right) > 0 \tag{6}$$

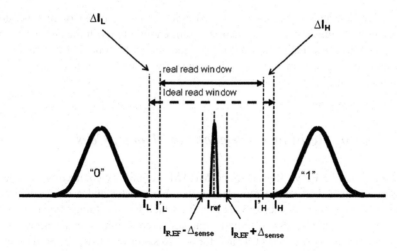

Fig. 6 Read window budget

And remembering (3), we can write:

$$I_{ref} - I_L > \Delta_{sense} + \Delta I_L$$
$$I_H - I_{ref} > \Delta_{sense} + \Delta I_H \qquad (7)$$

Summing these two relationships, we obtain:

$$I_H - I_L > 2\Delta_{sense} + \Delta I_H + \Delta I_L \qquad (8)$$

In other words, this is the minimum window that the process must be able to sustain in order to have a reliable reading, but, on the other hand, a constant design effort must be made to improve comparator sensitivity and to reduce reference-level spread.

Another problem for a correct placement of reference level in the reading window is the common mode shift in the current values, due to process parameter variations. Due to this, the reference level may be off-center with respect to the reading window, and, as a result, the "0" or "1" reading would be improperly influenced. To avoid this problem, the generation of the reference level (current in our example) should be made starting from a memory cell in such a way that tracks process variation.

Once the distributions of low and high currents in the cell have been well characterized, then a suitable choice of verify limits I_H and I_L is very important to optimize the yield of the memory device because state of art technology, in particular low maturity of emerging memory technologies and bigger and bigger memory size, don't allow high margins in the read window. If the limits are set too high compared with memory cell's performance, a low yield in programming will result, but a read window that is too small will lead to low yield during read.

Considering state of art emerging memories, a large spread of read-window width can be found. For some of them, it is not enough to reach the target bit error rate of a device or to maintain it during the operational lifetime. For this reason, in particular for big size memories, it is mandatory to embed in the chip an error correction engine.

3.1 Sensing Resistance Variations in Memory Cells

Memory cells like PCM, ReRAM, and MRAM are built by connecting in series a memory element that can be modeled as variable resistor and a device which is ideally a switch called a selector (see Fig. 7), which has no storage function but is necessary to enable access to a particular cell among others in an array; these memory cells are also called (1 Transistor + 1 Resistor or 1T1R). A similar cell is used in DRAM where the resistor is replaced by a capacitor (1 Transistor + 1 Capacitor or 1T1C), and, instead, flash are based on a different concept in which the selector and memory elements are collapsed into a single device (floating gate MOS).

The problem of sensing is to discriminate between "high" or "low" resistance values, associating with them two logic values stored in the memory cell. Due to the similarities among different technologies, it is justified to make a common analysis of the sensing problem, differentiating only the points peculiar to a specific technology. FeRAM instead doesn't work on resistance change, but on a charge sharing principle, and the sensing technique is more similar to DRAM.

A first method to evaluate the resistance value of memory cell is to inject into the cell through the bit-line a fixed (reference) current, reaching a voltage directly proportional to the resistance as stated by Ohm's law (9)

$$V_{BL} = R_{cell} I_{ref} \tag{9}$$

Then, V_{BL} can be easily compared with a reference voltage to determine if the resistances have to be considered "high" or "low". While the reference current is

Fig. 7 Cell-bias voltage can be applied only to the series of memory element and selector device

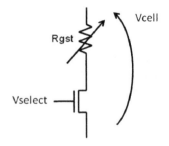

really applied to the resistance of Fig. 7, the resulting voltage that can be measured is the voltage across the series of the variable resistor (memory element) and the selector device which is measured at one end of the bit-line to which the cell is connected, because the intermediate connection point is not accessible from the external one, so we actually evaluate the sum of "memory" resistance R_{cell} and resistance of the selector device R_{on} (as the read current is usually low we discard the effects of parasitic resistance of the bit-line).

$$V_{BL} = (R_{cell} + R_{on})I_{ref} \tag{10}$$

It is mandatory that R_{on} is much less than the "low" value of the memory resistance to avoid reducing the signal available for reading, in particular, when the difference between R_{high} and R_{low} is relatively small, as, for example, in STT-MRAM where R_{high} is about 2 times R_{low}. The requirement about R_{on} leads to a demanding constraint on selector geometry. If it is an MOS device, it is not easy to make it wider because of the area occupation and, on the other hand, making transistor length smaller produces more leakage in the memory array. Generally, a trade-off between so many different constraints is needed.

A concept schematic of this sensing method is depicted in Fig. 8, the reference voltage shown in the figure is chosen as the product of I_{ref} and an intermediate resistance, midway between R_{high} and R_{low}, taken as a reference resistance.

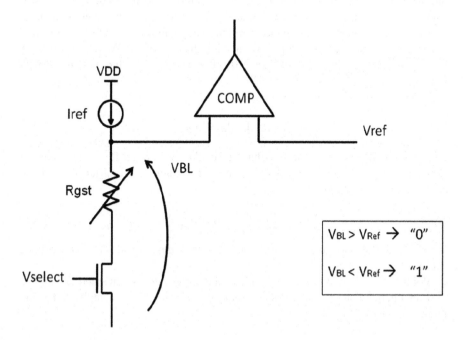

Fig. 8 Detecting resistance variation

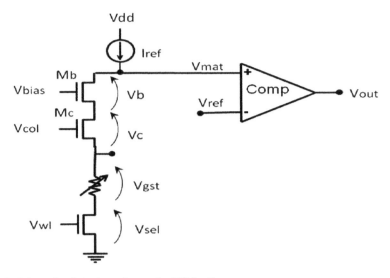

Fig. 9 Schematic of sensing scheme of a PCM cell

The circuit configuration shown in Fig. 8 has the drawback that the bias voltage across the memory element can reach dangerous levels, causing a read disturbance or, in case of STT-MRAM where memory element is a tunnel junction with a very thin insulator, breakdown of the junction. A solution is to interpose a bias device which has the role to clamp the voltage across the memory material, and, in addition, it decouples sensing node from the heavy capacitive load of the bit-line.

We can see this solution in Fig. 9, applied in the case of a PCM cell, where GST is schematized as a variable resistor [10]. The maximum voltage on the cell is limited to V_{bias} from the transistor Mb.

Voltage across the cell is given by:

$$V_{cell} = R_{gst}I_{ref} + V_{sel}. \qquad (11)$$

where V_{sel} is the drain to source voltage of cell's selector and R_{gst} is the value of Chalcogenide resistance; the voltage applied at the non-inverting terminal of the comparator (V_{mat}) is then:

$$V_{mat} = V_b + V_c + R_{gst}I_{ref} + V_{sel}. \qquad (12)$$

where V_c is the voltage across the column switch device and V_b is the voltage across M_b.

Voltage V_{ref} applied to the other side of the comparator is generated driving I_{ref} into an intermediate resistance R_{ref}, while V_c, V_b and V_{sel} are replicated using dummy devices

$$V_{ref} = V_b + V_c + I_{ref}R_{ref} + V_{sel} \qquad (13)$$

When the resistance of the cell is equal to R_{ref} the voltage at the sense node, V_{mat} is equal to V_{ref} and comparator is at trip point.

For PCMs, the read voltage across the GST should be somewhat below the hold voltage, (in the range of 0.4 V), but, actually, even voltages far from write range of the cell can cause read disturbance problems due to the high number of read cycles in the memory life. So the GST bias voltage should be the minimum value compatible with the development of enough current to ensure high enough reading speed. Additional considerations have to be made if the selector is a BJT; in this case, bit-line biasing has to take into account Vbe variation of the BJT, and resistive drop on the selected word-line, and Ron of the final word-line driver must be considered as depicted in Fig. 10, in this case, V_{mat} is given by:

$$V_{mat} = V_b + V_c + R_{gst}I_{ref} + V_{eb} + R_{wl}I_{ref}(1/\beta + 1) + V_r \qquad (14)$$

where V_b is the drain to source voltage of cascode M_b, V_c is the drain to source voltage of column selector, R_{gst} is the resistance of the memory element, V_{eb} is the emitter-base voltage of the selector (sel), R_{wl} is the parasitic resistance of word-line, β is the gain of BJT and V_r is the drain to source voltage of row decoder pull-down.

As we know, read voltage should be on the order of 0.1–0.2 V to ensure enough reading current and, at the same time, to avoid read disturbance problems. Column selector should be dimensioned for write operation, so V_c can be neglected.

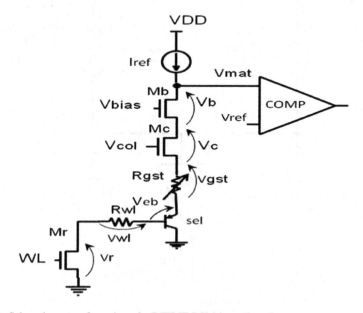

Fig. 10 Selected row configuration of a BJTBJT-PCM in read mode

V_r depends on the current contribution coming from all the cells of the same word-line that are read simultaneously; if many cells are read in parallel, this contribution can become not negligible. We have also to consider that, moving along the word-lines from the row decoder side on, the voltage drop due to word-line resistance increases, because the total resistance increases but also because current contributions coming from the cells read in parallel become important. We have seen already that the current flowing into the base of the BJT and then into the word-line is:

$$I_b = I_e/(\beta + 1) \tag{15}$$

In conclusion, it turns out that the configuration of the cell with a BJT selector places a maximum on the number of cells that can be read in parallel within a sub-array (tile) to limit the disturbance due to the V_{eb} reduction: the configuration with a majority of "one" is the worst case.

From (13) we see that V_{ref} depends on a reference resistor R_{ref} that ideally should have a value midway between high-resistance (R_{hi}) and low-resistance (R_{lo}) state of the memory. Sometimes this is achieved by setting the average between a R_{hi} cell and a R_{lo} cell with a dedicated circuit to track cell variations due to process,

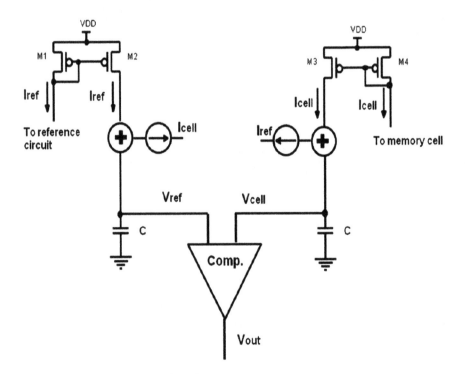

Fig. 11 Current comparison sensing scheme

geometry, voltage, and temperature. Setting of the two reference cells is made at the time of the wafer sort.

This sense concept is valid not only for PCM [11–13] but also for other resistive memories like ReRAM and MRAM [14–16].

A different sensing architecture [17] is shown in Fig. 11 in which I_{cell} and I_{ref} are cross-mirrored at ref-side and mat-side terminals of a voltage comparator, generating:

$$I_{matside} = I_{cell} - I_{refside}$$
$$I_{refside} = I_{ref} - I_{cell} \quad (16)$$

A current-to-voltage conversion is done through linear charging of the two capacitors C_M and C_R.

This schematic makes possible rejection of disturbances due to capacitive coupling with the substrate, power supply, and ground.

4 Sensing in STT-MRAM

Although the reading parameter of STT-MRAM is a variable resistance as in the previous cases, the available signal is lower compared with PCRAM, CBRAM, and oxide ReRAM, because, in spite of the fact that much effort is being made to radically improve this performance, today R_{high} is only about two times R_{low}, compared to a much higher ratio on PCRAM or ReRAM.

Due to these factors, much research has been directed, even more than in other emerging technologies, to find alternative sensing schemes able to achieve good performance with reduced signals [18].

Basic sensing configuration currently is still that shown in Fig. 8. With respect to the case already discussed for this scheme, a new consideration must be done in the case of a STT-MRAM cell, because even during read, it is possible to reach a voltage and current high enough to switch the cell. Considering the circuit of Fig. 8, we can write:

$$I_{ref}\left(R_{low} + R_{on}\right)_{max} < V_{ref} < I_{ref}\left(R_{high} + R_{on}\right)_{min} \quad (17)$$

$$I_{ref}\left(R_{high} + R_{on}\right) < V_{crit} \quad (18)$$

where $(R_{low} + R_{on})_{max}$ is the maximum value of the sum of MTJ low resistance and selector resistance and $(R_{high} + R_{on})_{min}$ is the minimum value of the sum of MTJ high resistance and selector resistance and V_{crit} is the voltage needed to switch the free layer ($V_{crit} = I_c R_{mtj}$) where I_c is the critical current to switch MTJ.

Considering resistance spread and the previously mentioned small difference between R_{high} and R_{low}, it is difficult to satisfy (17) for a large array and R_{on} becomes a critical point because it may be comparable both with R_{high} and R_{low}, further reducing the signal. The characteristic of selector is then critical in STT-MRAM cell.

The condition imposed by (18) states that voltage across the MTJ caused by I_{ref} must remain at a level far enough from the critical voltage needed to flip magnetization on the free layer. Positioning of reference current is then very important to avoid read disturbance as much as possible and at the same time to have the maximum possible signal. As a consequence a critical point for STT-RAM cell reading is the generation of the reference current that must be able to track cell-resistance variations due to process, temperature, and operating conditions and those coming from the dimensional spread inside the array. A possibility is given by using a high-resistance cell connected in parallel with a low-resistance cell and by generating I_{ref} as a function of the average of these currents:

$$I_{ref} = f\left(I_{R_{high}} + I_{R_{low}}/2\right) \tag{19}$$

Looking at the schematization of R/I characteristic of the MTJ shown in Chap.5, we can see that the available signal for reading, which is equal to $R_{high} - R_{low}$, is a decreasing function of the current flowing into the MTJ. At critical current (I_c), this difference is about 1/2 of its value at zero current.

Then, choosing a value for the reference current is a compromise between a higher signal window and a higher current, which means lower read latency but also less margin from MTJ switching point (Fig. 12). In spite of the fact that sensing method of Fig. 8 achieves the best performance in read latency, the achievable error rate can be quite high due to the large variability of the cell's parameters and the

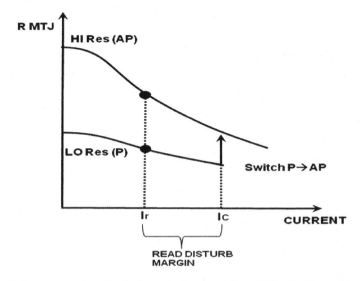

Fig. 12 Reference current positioning must be far enough from MTJ switching point

Fig. 13 Direction of current shift depending on MTJ resistance. Current flow must be on the parallelizing side to avoid disturbance

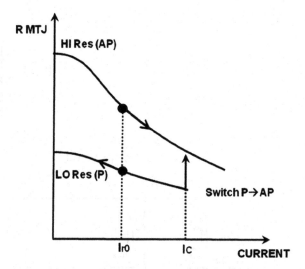

reduced window available for reading, for this reason, more complex reading circuits have been proposed for an optimized utilization of the read window.

A negative-resistance read scheme has been proposed [19] to increase reading signal remaining safe from spurious switching. The concept of this method is to include a feedback in the read current of the MTJ junction in such a way that starting from an initial reference current, if the MTJ is in a high-resistance state it will move reference to higher and higher currents, while if the MTJ is in a low-resistance state it will move reference to a a very small current, (see Fig. 13). Then a large read signal is obtained. This scheme can work without causing unwanted MTJ switching if the current direction in the MTJ is on the anti-parallelizing side, in which case the current increase in the high-resistance MTJ will reconfirm the data.

To overcome bit-to-bit variability affecting reading operation, a self-reference sensing method has also been proposed [20]. This is able to solve the problem of resistance spread because every cell act also as a reference for itself, and for a single cell it is always verified $R_{high} > R_{low}$.

Let's look at a concept scheme for this in Fig. 14 .

The read operation with this scheme needs multiple steps:

First read: A read current I_{ref_1} is applied to generate bit-line (bl) voltage Vbl1, which is stored in a capacitor C1. Vbl1 can be Vbl1l or Vbl1h, the bl voltages when the MTJ is at the low-resistance state or the high-resistance state, respectively.

Erase: Data "0" is written into the STT-RAM memory cell;

Second read: Another read current $I_{ref_2} > I_{ref_1}$ is applied to generate bl voltage Vbl2, which is stored in capacitor C2. Here I_{ref_2} is chosen in such a way that:

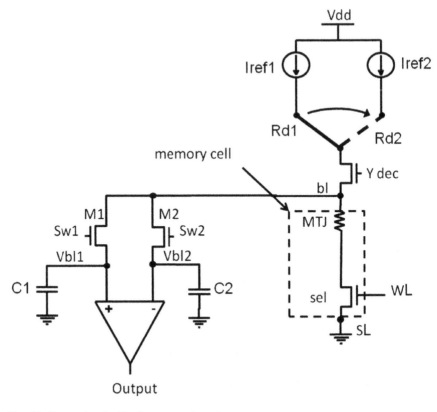

Fig. 14 Conventional self-reference sensing scheme

$$Vbl1l < Vbl2 < Vbl1h \tag{20}$$

The initial value of a STT-RAM bit can be readout by comparing Vbl2 and Vbl1.

Write back: Write the initial data back into the STT-RAM bit.

While the self-reference read method ensures a robust data reading, even with reduced the read window, it has a number of drawbacks:

- It requires two write operations in the worst case. If the original value of the cell was a "0", it is possible to avoid write back, remaining with only one write operation.
- When power supply fluctuations occur during the sensing process, the stored data in the STT-RAM bit may be lost.
- The introduction of a write operation to be performed during every read is an issue for endurance.

As we have seen, the most important issues of the self-reference scheme are related to the slow access time due to restore operation and the reliability issues coming from the need to perform a write at every read access. Trying to solve both of these drawbacks, a nondestructive, self-reference, sensing scheme was proposed [21] by leveraging the different slopes of the high and the low-resistance states with current (see Fig. 12): while R_{low} changes very smoothly with current, R_{high} changes very rapidly. Then, if it is possible to measure the slope of variation of a cell being probed, it can be decided if it is in R_{high} or R_{low} condition. This can be made by performing two read operations at different current levels and then comparing the results. Δ Vbl > Δ V_{ref} will indicate a fast resistance change, so then the cell is in a high-resistance condition; instead, Δ Vbl < Δ V_{ref} indicates a smooth variation typical of low-resistance condition.

Although this technique significantly reduces read latency and power consumption by eliminating the two write steps, the corresponding sense margin is smaller than that of the conventional self-reference; in addition, they are sensitive to selector/CMOS variations.

In conclusion, the quest for an optimum sense method for an STT-Ram cell is still in progress, and it is one of the key points for boosting further the development of this technology [18, 22].

5 Sensing in Ferroelectric Memories

Unlike the memory technologies discussed so far, the FeRAM reading method is based on charge sharing in a similar way to conventional DRAM, but exploiting the effect of polarization of the material which makes possible a stronger read signal.

Let's remember the configuration of the 1T-1C cell, in which the ferroelectric capacitor C_f is the memory element (see Fig. 15).

To read the cell it is necessary to exploit the ferroelectric effect of capacitor: after precharging the bit line at 0 V and activating pass transistor through WL line, PL is raised at Vdd (see the timing diagram of Fig. 16), establishing a capacitive divider

Fig. 15 Conventional FeRAM cell

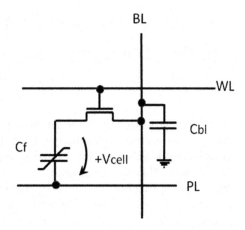

Fig. 16 FeRAM read access
timing

Fig. 17 Ferroelectric
capacitance is a function of
stored data

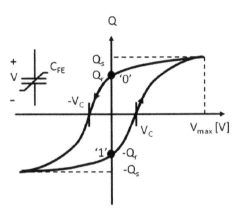

between C_f and bit-line capacitance C_{bl}. The charge variation of C_f resulting from
the voltage change depends on the initial state of the capacitor and on the amplitude
of voltage pulse as can be appreciated in Fig. 17. If the initial state was the "1" and
the pulse amplitude is wide enough to induce polarization switch, the total charge
moved is $2Q_r$ which can be modeled as a capacitance C_1. On the contrary, if the
initial state of the capacitor was "0", no polarization switch takes place and the
moved charge is only that coming from the displacement component which we can
model with a capacitance C_0. The result is that $C_1 > C_0$.

Therefore, the voltage developed on the bit-line can be one of the two values:

$$V_0 = C_0/(C_0 + C_{bl})V_{cell} \tag{21}$$

if the stored data is a zero, or

$$V_1 = C_1/(C_1 + C_{bl})V_{cell} \tag{22}$$

if the stored data is a "1".

This voltage difference can be detected by the sense amplifier which in principle
can be the cross coupled pair used for DRAM that is activated when the read signal
has been developed on the bit-line: positive feedback of cross-coupled pair drives
the bit-line to Vdd or gnd depending on the sign of the variation determined by
charge sharing (see Fig. 16) . It is important to underline that the read operation is
destructive, because, if the ferroelectric capacitor is initially in the state "1", the

application of a voltage pulse forces it to switch to the state "0". In a similar way to DRAM, it is then necessary to restore the original data stored in the ferroelectric capacitor after the reading. The cross-coupled inverter positive feedback can also automatically restore the data in the cell.

It should be noticed that data refresh in FeRAM is necessary only as a consequence of destructive reading, and ideally it is not required to maintain data a long time because FeRAM is a non-volatile memory. This is one of the most interesting features of FeRAM as compared with DRAM, because it cuts the big contribution of refresh to the needed standby power of DRAM memories.

The signal available from FeRAM is about two times the signal coming from an equivalent DRAM capacitor, but, to implement a correct reading, a reference voltage must be generated midway between V_0 and V_1.

At this point, the question comes up about how to design the reference voltage in such a way that a reliable reading of the memory cell is possible. The distribution of values of V_0 and V_1 depends on C_{bl} which is affected by process parameters related to the CMOS platform, process or the specific memory cell process and on C_0 and C_1 that depend on the characteristics of ferroelectric capacitor that change across the memory array. In addition other specific phenomena occurring in ferroelectric capacitors and already described in Chap. 4 contribute to V_0 and V_1 variability: "fatigue" degrades faster the cells that are accessed more often than the less-accessed cells, and "Imprint" which results in a voltage offset in both V_0 and V_1, This all implies that during read no fixed value of reference voltage (V_{ref}) can be used across the chip, but rather a variable reference voltage is required, to accurately track the process variation and the ferroelectric material degradation. In addition, temperature and voltage variations can impact V_{ref} in a different way, causing asymmetry in the reading signal fed to sense amplifier.

A conventional approach for generating a reference voltage for a column of memory cells is shown in Fig. 18.

The reference elements used to build the reference voltage are cells (one reference cell per bit-line) with their dedicated reference word-lines (Wl_{ref_i} and Wl_{ref_j}) running through the array. The array structure shown in the figure is similar to a checkerboard, so that, along Wli, there are memory cells only in "even" bit-lines and along Wlj there are cells only in "odd" bit-lines. Then, Wl_{ref_i} controls access to the row of reference cells that are connected to "odd" bit-lines and, similarly, Wl_{ref_j} controls access to the row of reference cells that are connected to "even" bit-lines. In this way, to every wordline activating a cell on a side of sense-amplifier corresponds a reference word-line activating a reference cell on the other side of the sense amplifier. C_{ref}, the reference capacitor of reference cells, always stores a "0", but it is sized larger than the corresponding memory capacitor C_0, so that the signal developed (V_{ref}) is midway between V_1 and V_0.

The timing diagram used for a read operation is shown in Fig. 19.

The bit-lines and the storage nodes of the reference cells, are precharged to 0 V prior to a read operation, Wl_i and Wl_{ref_i} are activated together and later a simultaneous step voltage is applied on the PL_i and PL_{ref_i}. Due to the capacitive ratio

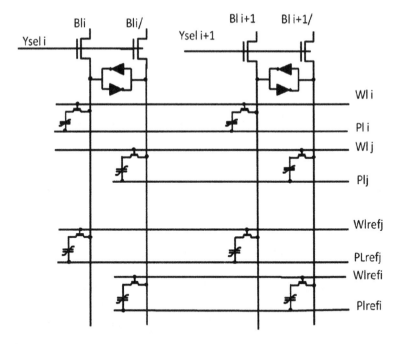

Fig. 18 Reference structure with one capacitor per column

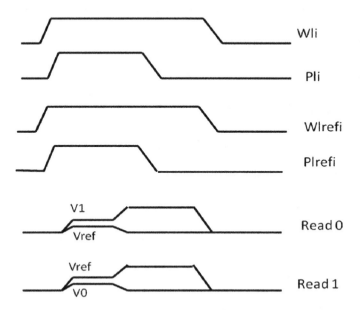

Fig. 19 Timing diagram of read operation using 1C per bit-line reference

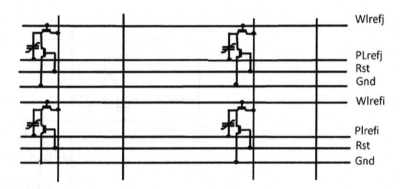

Fig. 20 Complete reference architecture using one capacitor

C_{ref}/C_{bl}, the $Bl_i/$ is raised to V_{ref}, while Bl_i is raised to either V_0 or V_1 depending on the stored data value.

Then the activation of the sense amplifier allows the comparison with V_{ref} and sends the bit-line of higher voltage (BL or BL/) to Vdd and the bit-line of lower voltage to 0 V (see Fig. 19).

The storage nodes of the reference cell are pulled to ground before PL_{ref_i} through a reset transistor added to each reference cell. This guarantees that C_{ref} is never allowed to switch whatever is the sensed data.

In Fig. 20, is shown a detail of the reference structure with the reset transistor driving to GND the storage node of C_{ref}. The operation is driven by a signal RST connected to the gates of reset transistors.

Another methodology [23] consists of using two capacitors C_{ref_0} and C_{ref_1}, each with two access transistors controlled by different signals, as shown in Fig. 21, instead of a single oversized capacitor. The capacitors are half the size of a memory-cell capacitor, with C_{ref_0} always storing a "0" and C_{ref_1} always storing a "1". Therefore, if they are accessed simultaneously by raising RWL, RPL_0, and RPL_1, they generate a reference voltage given by (23), where and C_0 and C_1 are the approximate ferroelectric capacitances related to "0" and "1" switching. A precharge phase before next reading is needed to restore C_{ref_0}, C_{ref_1} at "0" and "1" respectively, this is accomplished by rising PCH_1, PCH_2 and RPL_0 simultaneously with RPL_1 and RWL at GND.

$$V_{ref} = V_{dd}(C_0/2 + C_1/2)/(C_0/2 + C_1/2 + C_{bl}) \qquad (23)$$

A drawback of this reference scheme is that it fatigues the reference cells faster than the memory cells by accessing the reference row each time a memory row is accessed. Moreover, permanently writing a "0" into a C_{ref_0} and a "1" into C_{ref_1} can cause imprint in these capacitors, but this problem can be avoided by exchanging the data between C_{ref_0} and C_{ref_1} whenever they are accessed.

Both the reference schemes discussed so far include a row of reference cells in the array which is accessed by a separate word-line (RWL) and plate-line (RPL).

Fig. 21 Reference architecture using two capacitors

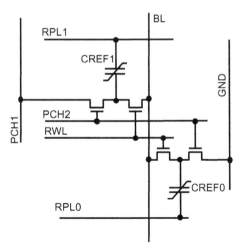

A problem of these schemes is that reference cells are always accessed for each row access and then they are much more fatigued than array cells. For example, sequentially accessing n rows of an array fatigues the reference cell n times faster than each individual memory cell.

A possible solution of this drawback has been proposed [24] in which a reference cell is associated to each row of the array, so that a cell is fatigued only when his row is accessed, however this leads to a remarkable increase of the number of reference cells.

References

1. "Circuit Design in Emerging Technologies" TD forum, ISSCC 2006
2. "Non-Volatile Memories Technology and Design", Memory Design Forum, ISSCC 2004
3. "Non volatile memory circuit, design and technology" Short course F1, ISSCC 2007
4. G. Campardo et al. "An Overview of Flash Architectural Developments" Proceeding of the IEEE, April 2003
5. R. Micheloni et al. "The Flash Memory Read Path: building blocks and critical aspects" Proceeding of the IEEE, April 2003
6. G. Campardo et al. "VLSI-Design of Non-Volatile Memories", Springer Series in Advanced Microelectronics, 2005.
7. "DRAM's in the 21st Century" IEDM1996 short course, organizer: S. Shinozaki
8. "Advanced Dynamic Memory Design" Memory Design Forum, ISSCC 2005
9. G. Campardo et al. "Architecture of non volatile memory with multi-bit cells" Elsevier Science, Microelectronic Engineering, Volume 59, Issue 1–4, November 2001, pp. 173–181
10. G. Casagrande "Phase Change Memory" Tutorial, ISSCC 2004
11. H. Chung et al. "A 58 nm 1.8 V 1 Gb PRAM with 6.4 MB/s Program BW" ISSCC 2011, Digest of technical papers
12. C. Villa et al. "a 45 nm 1 Gb 1.8 V phase-change memory" ISSCC2010, Digest of technical papers

13. Y. Choi et al "A 20 nm 1.8 V 8 Gb PRAM with 40 MB/s Program bandwidth" ISSCC2012, Digest of technical papers
14. S. Hollmer "A CMOS Compatible Embedded 1 Mb CBRAM NVM" ISSCC 2012, Digest of technical papers
15. W. Otsuka "A 4 Mb Conductive Bridge Resistive Memory with 2.3 GB/s Read-throughput and 216 MB/s Program throughput" ISSCC2011, Digest of technical papers
16. R. Fackenthal et al. "A 16 Gb Re-RAM with 600 MB/s write and 1 GB/s read in 27 nm technology" ISSCC2014, Digest of technical papers
17. F. Bedeschi et al. "A multilevel cell bipolar selected Phase Change memory" ISSCC2008, Digest of technical papers
18. T. Kawahara et al. "Spin-transfer torque RAM technology: review and prospect" Microelectronic reliability 52 (2012), pp. 613–627
19. D. Halupka et al. "Negative resistance Read&Write schemes for STT-MRAM in 0.13 um CMOS" ISSCC 2010 Digest of technical papers
20. G. Jeong et al. "A 0.24-μm 2.0 V 1T1MTJ 16 Kb non-volatile magnetoresistance RAM with self-reference scheme" IEEE J. Solid-State Circuits, vol. 38, no. 11, pp. 1906–1910, Nov. 2003
21. Z. Sun et al. "Voltage driven Non Destructive Self-Reference Sensing Scheme of Spin Transfer Torque Memory" IEEE trans. on VLSI Systems vol. 20, no. 11, Nov. 2012
22. R. Takemura et al. "highly scalable Disruptive Reading scheme for Gb-scale SPRAM and beyond" ISSCC2010, Digest of technical papers
23. A. Sheikholeslami et al. "A survey of Circuit Innovation in Ferroelectric Random-access Memories" Proc. of the IEEE, vol. 88, no. 5, May 2000
24. A.G. Papaliolios "Dynamic Adjusting Reference Voltage for Ferroelectric Circuits" U.S. Patent 5218566, June 8, 1993

Algorithms to Survive: Programming Operation in Non-Volatile Memories

Alessandro Cabrini, Andrea Fantini and Guido Torelli

1 Why Algorithms to Write and Erase Non-volatile Memories?

Non-volatile memories offer the possibility to maintain the stored digital information over years without a power supply. This requires programming mechanisms that physically modify the memory cell: this physical alteration must be "hard" in order to guarantee the non-volatility of the stored information. We could say that, in general, the harder it is to intentionally modify the cell, the better is the data retention over time.

When programming a non-volatile memory, we modify a physical parameter of the cell in a hard but reversible way. The chosen programmable parameter must be able to control the electrical behavior of the cell, as is required during a read operation, when the "state" of the cell is electrically sensed. For instance, in the case of flash memory, programming consists of trapping electrons in the floating gate of the memory cell, whereas, in phase-change memory (PCM), writing is achieved by modifying the phase (amorphous or crystalline) of a suitable phase-change material. The programmable parameter of a flash memory cell is thus the amount of charges trapped in the floating gate, which controls the cell threshold voltage, whereas, in the case of PCM, it is represented by the relative amount of one phase (e.g., the

A. Cabrini (✉) · G. Torelli
Department of Electrical, Computer and Biomedical Engineering, University of Pavia, Pavia, Italy
e-mail: alessandro.cabrini@unipv.it

G. Torelli
e-mail: guido.torelli@unipv.it

A. Fantini
imec, Leuven, Belgium
e-mail: andrea.fantini@imec.be

© Springer International Publishing AG 2017
R. Gastaldi and G. Campardo (eds.), *In Search of the Next Memory*,
DOI 10.1007/978-3-319-47724-4_7

crystalline phase) with respect to the other (e.g., the amorphous phase), which controls the electrical resistance of the cell.

In general, the programmable parameter can be varied over a given continuous range from a minimum to a maximum value. This means that programming substantially corresponds to an analog trimming operation. This "trimming" obviously requires accuracy, but also intrinsically offers the possibility to implement multi-level programming (i.e., the possibility to store more than one bit in a single cell). The main problem is that, due to variability in the fabrication process as well as in operating and environmental conditions, various different cells will react to specific programming procedures in different ways. In order to ensure adequate programming accuracy for all the cells in the array, it is therefore necessary to devise specific programming algorithms. An adequate programming algorithm should also take other parameters into consideration, including programming time as well as reliability issues (in this respect, it is worth to recall that non-volatile memory programming is a hard operation and, hence, the materials of the cells undergo significant stresses). A number of algorithms have been developed for this purpose, each algorithm being targeted to a specific kind of non-volatile memory, taking advantage of the specific features and programming mechanisms of the targeted memory type and having specific goals.

In this scenario, the most-used programming algorithms for a number of non-volatile memories (including flash memories [1, 2] and PCMs [3, 4]) are based on the program-and-verify (P&V) approach. Following this approach, the program operation is divided in partial programming steps. At the end of each step, the cell is verified (essentially, the cell is compared with a suitable reference). If the cell has reached the required state, the program operation over this cell stops; otherwise, the algorithm continues until the target state is reached or a given number of unsuccessful programming steps have been performed (in the latter case, the cell is considered as failed).

A key issue of any P&V algorithm is the intrinsic trade-off between resolution and overall program time. High programming resolution is beneficial for increasing the spacing between adjacent programmed distributions, which leads to better read margins and, hence, ensures fast and robust read signals. Unfortunately, improving the resolution of the P&V algorithm determines an increase in the overall duration of the program operation. Indeed, if we assume that the programmable parameter P_P has to be varied by an amount ΔP_P, the number N of P&V steps required to span the range ΔP_P can be expressed as:

$$N = \frac{\Delta P_P}{\delta P_p} \tag{1}$$

where δP_P is the variation of P_P obtained at the end of a single P&V step. The time T_P necessary for a complete programming is:

$$T_P = Nt_{P\&V} \tag{2}$$

where $t_{P\&V}$ represents the time duration of a single P&V step. Since the programming resolution R can be expressed in terms of equivalent bits[1] as:

$$R = \log_2 N \tag{3}$$

T_P can be rewritten as:

$$T_p = 2^R \cdot t_{P\&V} \tag{4}$$

which shows that the better the required resolution R, the higher the necessary programming time T_P.

2 Introduction to Flash Algorithms

A number of programming algorithms have been developed for flash memories [6–8], targeted at improving both performance and reliability. The most popular algorithms include:

- program-and-verify: as mentioned above, a program operation consists of a number of partial programming steps, each followed by a read step, whose result enables or not an additional program step (program-and-verify is performed at the cell level, since each cell can be independently programmed) [9–11];
- erase-and-verify: an erase operation is followed by a verify operation, whose result enables or not an additional erase operation (erase-and-verify is obviously performed at the block level);
- soft programming after erasing: the cells whose threshold voltage has been driven negative in the previous erasing operation, are re-programmed until they reach a positive threshold voltage (this operation is only required in NOR-type flash memories) [7];
- program-all-cells, erase-all-cells, program-required-cells: this way, the same erase-program history and, hence, the same ageing is provided to all cells in the memory block, which also makes it easier to ensure the best endurance [9, 10];
- staircase up (or incremental step) programming: the successive applied program pulses have an increasing amplitude (this approach ensures the best trade-off between programming time and accuracy) [5, 12, 13];
- staircase up erasing: an increasing staircase voltage is applied for erasing [7];
- predefined programming sequence within a string in NAND-type flash memories: programming starts from the cell closest to the source selector and proceeds

---

[1]The achievable resolution is in practice worse due to non-idealities in program and verify operations [5].

upwards along the string (this approach prevents problems due to the background pattern dependence: the threshold voltage of a cell as seen by its control gate depends on the programmed state of the lower cells in the string) [8].

Most of the above algorithms are specific to flash memories, and therefore will not be addressed here. In this section, we focus our attention on the program-and-verify algorithm, which is also used for other kinds of non-volatile memories (including PCMs and ReRAMs), where the "cumulative" effect of the programming operation is exploited (substantially, the effects of successive applied pulses sum up until the target state of the cell has been reached).

Considering both NAND and NOR array organization, in most recent generation of flash memories, multi-level storage (ML) [14–18] represents the standard approach. Following the ML approach, a memory cell can be programmed to any of $m = 2^n$ different threshold levels or, when programming and reading non-idealities are taken into account, to any of 2^n different distributions of threshold values (see Fig. 1, where shaded areas correspond to read and reliability margins that must be provided between adjacent programmed levels). A single cell can therefore store as many as n bits. This makes it possible to overcome the traditional one-to-one relationship between cell count and memory capacity, thereby increasing storage density for any given fabrication technology generation.

Programming a flash memory cell is usually achieved by means of a sequence of partial program steps interleaved by verify steps (the aforementioned P&V approach). After each program step, the cell is sensed, in order to verify whether or not the cell threshold voltage V_T, which was initially set to the erased (low) level $V_{T, ER}$, has reached the target value $V_{T,P}$. If this is not the case, the program operation goes on with another partial program step. Finally, the program operation on the considered cell ends when V_T reaches the desired value. Thanks to the selectivity in

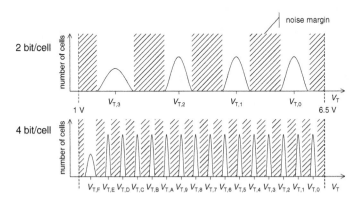

Fig. 1 Conceptual representation of programmed threshold voltage distributions for ML flash memory cells in the case of 2-bit/cell (i.e., 4-level) and 4-bit/cell (i.e., 16-level) storage. The subscript k in threshold voltage $V_{T,k}$ refers to the hexadecimal code of the corresponding state ($V_{T,3}$ in the upper plot and $V_{T,F}$ in the lower plot represent the erased state, referred to as $V_{T,ER}$ in th text)

addressing the array bit-lines and, hence, in providing the programming drain voltage to each individual cell in any selected row when required, the P&V algorithm can be simultaneously applied to all the cells in a word (NOR-array organization) or even in a page (NAND-array organization) ("bit-by-bit intelligent programming"), thus achieving adequate program throughput.

According to the most used ML programming technique, all programming steps have the same time duration and a constant cell-drain voltage. The voltage applied to the control gate is instead increased by a constant amount δV_{GP} at each step, thus resulting in a staircase waveform of the gate program voltage. With this approach, after an initial transient, a constant threshold shift $\delta V_{T,i} = V_{T,i} - V_{T,(i-1)}$ is achieved at each i-th partial program step, where $\delta V_{T,i} = \delta V_{GP}$ (this feature is common to both channel hot-electron injection and Fowler-Nordheim tunneling programming). Therefore, when using a staircase waveform for the gate voltage, the time slope of the total threshold voltage shift ΔV_T turns out to be identical to the slope of the applied gate voltage V_{GP}.

When using a P&V algorithm, the obtained V_T distribution width depends on the increment δV_{GP} of the applied gate voltage. Indeed, neglecting non-idealities such as sense circuit inaccuracy and voltage fluctuations, the last programming pulse applied to a cell causes its V_T to be shifted above the decision level by an amount at most as large as δV_{GP}. It is therefore evident that, to achieve high program accuracy, it is necessary to reduce δV_{GP} as much as possible. The programming resolution R can be expressed as:

$$R = log_2 \frac{\left(V_{T,P}^{max} - V_{T,ER} \right)}{\delta V_{GP}} \tag{5}$$

where $V_{T,P}^{max}$ is the uppermost threshold voltage level that can be programmed.

As the available voltage window $W_{VT} = V_{T,P}^{max} - V_{T,ER}$, where all V_T levels can be allocated (usually referred to as threshold window), can not be increased beyond a given limit due to reliability reasons, to increase the storage capacity of a cell by one bit, it is necessary to proportionally decrease δV_{GP}, assuming read and reliability margins are maintained proportional to the distribution widths. Indeed, from (5), we have $\delta V_{GP} = W_{VT}/2^R$.

As mentioned above, the time T_P needed to program a cell to the target level $V_{T,P}$, starting from the erased level $V_{T,ER}$, is inversely proportional to the program step δV_{GP}:

$$T_P = \sim \frac{\left(V_{T,P} - V_{T,ER} \right)}{\delta V_{GP}} t_{P\&V} = 2^R \frac{\left(V_{T,P} - V_{T,ER} \right)}{W_{VT}} t_{P\&V} \tag{6}$$

where $t_{P\&V}$ is the time duration of a single program step (including the time required to carry out program verify). To cope with this limitation, it is possible to use architectures with high write parallelism, i.e., to simultaneously program/verify a large number of cells. The main drawback of increasing parallelism is the need for a larger silicon area. Indeed, the high voltages required for programming and erasing are generated on chip by means of dedicated voltage elevators (generally based on the charge pump approach). These voltage elevators have a driving capability that is proportional to the occupied silicon area and, hence, increased parallelism (which determines a proportional increase in the current needed for program operations) implies that a larger silicon area must be allocated to implement the required high-voltage generators. The silicon area increase is not determined only by the augmented programming current budget. Indeed, the verify operation requires sensing circuits, and the degree of parallelism also determines the number of sensing circuits that must be included in the memory device, thus leading to a further need for silicon area. However, the increased storage density offered by the ML approach, largely compensates for the resultant silicon-area penalty.

3 Write Algorithms for PCMs

3.1 Introduction

The working principle of a PCM cell [3, 4] relies on the physical properties of chalcogenide materials (the most common being the $Ge_2Sb_2Te_5$, GST, alloy) that can be reversibly switched between two structural phases having significantly different electrical resistivity: the amorphous phase (highly resistive) and the (poly)crystalline phase (less resistive). A difference of about two orders of magnitude exists between the cell resistance in the SET and the RESET state, which easily enables non-volatile bi-level storage and can, in principle, be exploited for multi-level storage.

As depicted in Fig. 2a, which is referred to the case of the μ-trench architecture, a PCM cell is composed of a thin GST film, a resistive element named heater, and two metal electrodes, i.e., the top electrode contact (TEC) and the bottom electrode contact (BEC), which enable external electrical connection. The phase transition is obtained by applying suitable electrical (current or voltage) pulses to the GST alloy, which increase the temperature inside the GST (for instance, in the case of Fig. 2b, these pulses can be applied to the cell by controlling the gate voltage of transistor Y_0). Only a portion of the GST layer, which is located close to the GST-heater interface and is referred to as active GST, undergoes phase transition when the PCM cell is stimulated. During a RESET program operation, the active GST volume is heated above its melting point and then rapidly cooled down to the normal operating temperature ("fast quenching"). The cooling rate must be faster than the crystallization rate so as to freeze the atoms in a disordered structure, thus producing an amorphous volume (RESET state). To program the cell to the SET state, the active GST is generally heated to a temperature between a critical value for the crystallization process (referred to as crystallization temperature, T_x) and its melting point

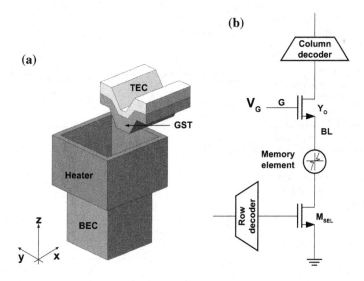

(b)

(a)

Fig. 2 PCM cell architecture (μ-trench cell): **a** physical structure of a PCM cell; **b** read/program path (the PCM cell is the memory element)

for a predetermined time interval. This way, the nucleation and the microcrystal growth inside the GST lead to a (poly)-crystalline active volume (SET state).[2]

In principle, the PCM cell can be programmed to any target resistance R_T between the SET and the RESET resistance, R_{SET} and R_{RST}. This can be obtained by following two complementary approaches: partial-SET programming [19] and partial-RESET programming [20, 21]. In the former approach, the cell is first brought into the RESET state (i.e., the GST is fully amorphized) and, then, a partial-SET programming pulse is applied so as to partially crystallize the active volume. In the latter case, the cell is first brought into the SET state (which means that the GST is now fully crystallized), and, then, a partial-RESET pulse is applied in order to partially amorphize the active volume. By applying suitable sequences of partial-SET (partial-RESET) programming pulses, each preceded by a full-RESET (full-SET) operation so as to re-initialize the cell, it is possible to change the cell resistance from R_{RST} to R_{SET} (or vice versa). The evolution of the cell resistance as a function of the applied pulse amplitude during programming in the two cases is compared in Fig. 3. In particular, programming with partial-SET pulses enables changing the cell resistance from A to B, whereas, in the case of partial-RESET programming, the resistance is changed from A' to C. Since the amorphization process is very fast when compared to crystallization, partial-RESET

---

[2]Another SET programming approach consists of heating the active GST up to its melting point and then cooling it down slowly to the normal operating temperature, so as to enable crystallization in an ordered structure. An implementation of this approach (staircase-down SET programming) will be discussed in Sect. 3.2.

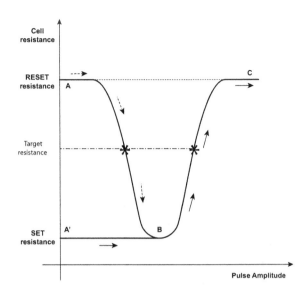

Fig. 3 Evolution of the cell resistance during a programming sequence as a function of the applied pulse amplitude (briefly, programming characteristic) in the case of partial-SET (*dashed arrows*) and partial-RESET (*continuous arrows*) pulses. A single pulse is applied to the cell which is initially either in state A (partial-SET programming) or in state A' (partial-RESET programming)

programming promises, in principle, shorter programming time than partial-SET programming. However, since a partial-RESET pulse must amorphize a portion of the active GST, higher amplitudes are typically required to raise the GST temperature above its melting point T_M, as apparent from Fig. 3.

It is worth noting that, for the same value of programmed resistance R_P, the two approaches determine two different distributions of the amorphous and the crystalline phase in the active GST volume. This leads to different behaviors of the cell in the two cases when considering, for example, the temperature coefficient of the cell resistance, the resistance drift (i.e., the increase in the programmed resistance over time) [22–24], or the impact of crystallization on reliability [25, 26].

As explained in the next sections, in order to guarantee an accurate control of the cell resistance, also in the case of PCM, it is necessary to implement programming sequences based on the P&V approach (which can be realized by following either partial-SET or partial-RESET programming).

3.2 Bi-level Programming

When considering bi-level programming, the cells in the array are alternatively programmed either to the crystalline (SET) or to the amorphous (RESET) state. The simplest way to implement bi-level PCM programming is to use two different programming pulses, as depicted in Fig. 4. On the one hand, as stated above, the amplitude of the RESET pulse, V_{RST}, must be large enough to raise the temperature of the active GST above T_M. The active GST is then amorphized provided that the falling edge of the RESET pulse is adequately sharp. The main problem is to

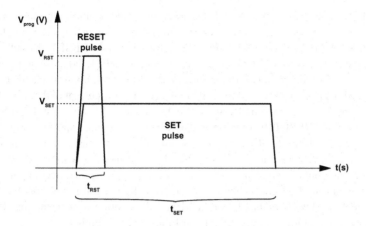

Fig. 4 Simple programming pulses for bi-level programming

determine an adequate value of V_{RST} so as to bring all the cells in the considered array into the RESET state. In fact, due to fabrication process spreads, different values of V_{RST} are required to melt the active GST in different cells. Unfortunately, excessively increasing the value of V_{RST} (as required to ensure the amorphization of all the cells by means of the RESET pulse) can determine reliability issues: indeed, cells that melt at lower values of V_{RST} will undergo a larger stress if programmed with high values of V_{RST}.

On the other hand, the SET pulse amplitude, V_{SET}, must not bring the active-GST up to its melting point, but must be large enough to adequately speed-up the crystallization process so as to obtain a fully crystalline state after applying a SET pulse with time duration t_{SET}. A trade-off between V_{SET} and t_{SET} should be chosen. Indeed, provided that the temperature inside the GST is raised above the (critical) crystallization temperature T_x, the higher the temperature, the shorter the required time duration of the SET pulse. Also in this case, the problem is to choose an adequate value of the SET programming voltage. Indeed, V_{SET} should be not excessively high to prevent the melting of the GST, and, hence, the amorphization of the active region (which would cause a SET programming failure), but should be as high as possible to adequately limit t_{SET}, thus increasing program throughput.

In practice, the simple use of a single SET pulse and a single RESET pulse with predetermined amplitude and duration to program a memory array is not feasible (due not only to process spreads, but also to the adverse effects caused by parasitic elements, such as, for example, the bit-line resistance, which impact on the actual shape of the programming pulse).

Let us analyze single-pulse programming in more detail. Assume first that all the cells in the array are in their RESET state and apply the same SET programming pulse to each cell in the whole array. Reading the current of the cells after applying the SET pulse enables determining the distribution of the SET read current, hereinafter simply referred to as SET-state distribution or SET distribution (the read operation must be performed with a direct memory access, which corresponds to an analog read operation of the cell current). Similarly, it is possible to obtain the distribution of the RESET read cell current (hereinafter referred to as RESET-state distribution or RESET distribution): assume now that the cells are initially programmed to their fully crystalline state and apply the same RESET pulse to all cells. Also in this case, reading the current of the cells enables determining the corresponding RESET-state distribution. What it is highly probable is that the SET and RESET distributions will (partially) overlap, as shown in Fig. 5a: the worst-case cell of the RESET-state distribution, RST_WC, has a programmed resistance lower (and, hence, a read cell current higher) than the worst-case cell of the SET-state distribution, SET_WC. The cells with a resistance in the overlapping region (shaded region in the figure) cannot be read correctly since it is not possible to discriminate if they belong to the SET or the RESET distribution. This means that the used programming algorithm is not adequate for bi-level programming.

In particular, the problem is that the used V_{RST} is not sufficiently high to fully amorphize the cells in the upper part of the RESET distribution (i.e., in the part of this distribution close to RST_WC), and the thermal energy delivered during the SET pulse is not sufficient to fully crystallize the cells belonging to the lower part of the SET distribution (i.e., to the part of this distribution close to SET_WC).

In order to obtain SET and RESET distributions separated by an adequate margin (as shown in Fig. 5b), modified programming pulses are typically used. In particular, to obtain a tight SET distribution, it is possible to use a staircase-down (SCD) SET programming pulse (Fig. 6a) [27] whereas, to achieve the RESET state, it is possible to use an N-pulse staircase-up RESET programming approach, where N pulses with increasing amplitude, each followed by a verify pulse, are applied (Fig. 6b).

The basic idea behind the SCD SET programming technique is to approximate a programming pulse that first brings the active GST volume above its melting point

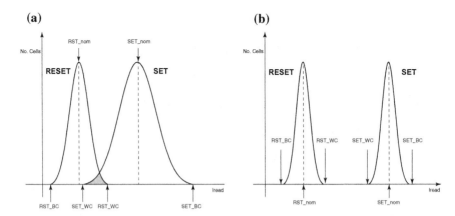

Fig. 5 Current distributions obtained after bi-level programming of the whole memory array corresponding to the cases of **a** overlapping distributions and **b** well-separated distributions (*Iread* = read cell current; *nom* = nominal; *BC* = best case; *WC* = worst case)

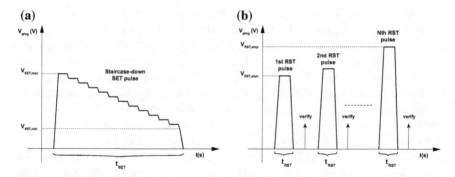

Fig. 6 Bi-level PCM programming algorithms: **a** Staircase-down SET programming; **b** N-pulse staircase-up RESET programming

and then slowly cools it down to room temperature ("slow quenching"), so as to give the atoms the time necessary to crystallize in an ordered lattice. To this end, a staircase starting from an adequately high $V_{SET,max}$ and decreasing to a given $V_{SET,min}$ with steps of constant amplitude-δV_{SCD} and constant time duration δt_{SCD}, is applied. The initial voltage $V_{SET,max}$ is chosen so as to melt the active GST, which is then slowly cooled thanks to the applied decreasing-staircase waveform. For an optimal crystallization of the GST, cooling should be as slow as possible. A key problem is that the SET programming time, t_{SET}, is inversely proportional to the slope, $S_{SCD} = \delta V_{SCD}/\delta t_{SCD}$, of the staircase waveform. In fact, since the number of steps, N_{SCD}, of the SCD pulse is given by:

$$N_{SCD} = \frac{\left(V_{SET,max} - V_{SET,min}\right)}{\delta V_{SCD}}, \qquad (7)$$

t_{SET} can be expressed as:

$$t_{SET} = \delta t_{SCD} N_{SCD} = \frac{\left(V_{SET,max} - V_{SET,min}\right)}{S_{SCD}}. \qquad (8)$$

The optimal value of S_{SCD} is thus a trade-off between accuracy and program throughput. It is worth to point out that, in contrast to the case of flash memory programming, the variation of the cell resistance due to each step of the applied SCD waveform is not directly proportional to δV_{SCD}. Indeed, the crystallization process, which is strongly dependent on the temperature in the active GST, is intrinsically random. This makes the programming operation of a PCM array totally different with respect to the case of a corresponding flash memory array. In particular, the crystallization speed in any specific cell of the array is not the same in the case of repeated programming operations, but may vary any time the same SCD pulse is applied and can randomly determine an incomplete crystallization at the end of the SCD pulse. For this reason, in practical algorithms, the cell is verified at the end of the SCD pulse and, in case the detected resistance is too high, the memory controller applies a further SCD pulse (in some cases, the algorithm can modify one or more of the parameters $V_{SET,max}$, $V_{SET,min}$, δV_{SCD}, and δt_{SCD}). Typically, the cell is considered as failed if its read current does not meet the required minimum value (SET_WC) after a predetermined number of SCD pulses.

N-pulse staircase-up RESET programming can be considered as an adaptive algorithm that aims at minimizing the stress on the PCM cell. To this end, the RESET procedure starts applying a first RESET pulse with amplitude $V_{RST,start}$ (high enough to melt the GST) and a duration on the order of tens of ns. After the first RESET pulse, the cell is verified: if its resistance is sufficiently high, the operation stops, otherwise a new RESET pulse with increased amplitude and the same time duration is applied and a verify operation follows. Since the temperature reached by the GST increases at each applied step, the produced volume of the amorphous phase will also increase, thus resulting in a higher value of the cell resistance. The above procedure is repeated up to N times, and, as in the case of the SCD SET programming algorithm, the cell is considered as failed if its resistance does not meet the target requirements after the last verify step.

Both SCD SET and N-pulse staircase-up RESET programming RESET programming are performed in parallel over a number of cells which is limited by the maximum current that the programming circuits can deliver (the parallelism must not necessarily be the same in the two cases).

At the end of a complete programming operation, the spacing between the RESET-state and the SET-state distribution (i.e., between RST_WC and SET_WC) will depend on the parameters used to implement the two programming algorithms and, in general, will be larger when using algorithms requiring a longer programming time.

3.3 Multi-level Programming

When considering a bi-level operation of a PCM cell, the programming algorithms must be able to switch the resistance from its minimum to its maximum value (points A and A', respectively, in Fig. 3). From this point of view, it is not important to control how the resistance evolves from the minimum to the maximum value (RESET operation), and vice versa (SET operation). The situation changes when multi-level programming [28–30] is addressed. As mentioned above, various approaches have been proposed to obtain a specific target resistance R_T. In particular, it is possible to reach R_T starting from either A or A'. The differences between the two cases depend on the phenomena involved.

By following the most common approach, referred to as partial-SET staircase-up (SCU) programming, the cell is initially brought to its high-resistance state through a RESET operation which amorphizes the active GST (first pulse in Fig. 7). After the initial amorphization of the active phase-change material, the algorithm activates the crystallization of the active GST thanks to a series of pulses with

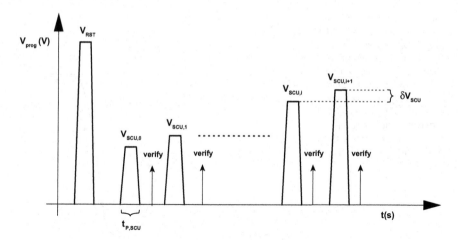

Fig. 7 Time diagram of a partial-SET staircase-up programming sequence. The cell is initially amorphized through a RESET pulse and, then, partial crystallization starts controlled by (partial-) SET programming pulses with increasing amplitude, fixed time duration, and fast quenching. Each programming pulse is followed by a verify operation

increasing amplitude (the amplitude increases by a constant amount δV_{SCU} at each step) and fixed time duration, $t_{P,SCU}$. Fast quenching is provided to each pulse.

Essentially, thanks to the temperature increase determined by the i-th SCU pulse, the crystal fraction in the active GST increases at each programming step and, hence, the effective resistance of the cell progressively decreases. In practice, there is a kind of cumulative effect on the GST that, by integrating the energy delivered at each programming pulse, determines the formation of the desired (poly)-crystalline phase.

As in the case of P&V algorithms used in flash memories, after each programming pulse, the state of the cell is verified and, if the cell has reached the target resistance R_T, the programming algorithm stops. With a suitable choice of parameters δV_{SCU} and $t_{P,SCU}$, it is possible to accurately vary the cell resistance from its maximum to its minimum value. It is worth noting that, in case the SCU programming sequence is not stopped once the target minimum resistance is reached, applying further pulses with increasing amplitude causes the cell resistance to increase. In fact, if the amplitude of an SCU pulse is sufficiently high to raise the GST temperature above the melting point, the active material will start to (partially) amorphize, which brings back the cell towards its high-resistance state. This represents the most critical issue when applying the (partial-)SET SCU algorithm.

By plotting the cell resistance measured after each programming pulse versus the pulse count, we obtain the so-called cumulative programming characteristic (Fig. 8), which is quite similar to the bell-shaped curve A-B-C (single-pulse programming characteristic) depicted in Fig. 3.

As mentioned above, in order to be able to span the whole resistance window, an adequate choice of parameters δV_{SCU} and $t_{P,SCU}$ is essential. In fact, the crystallization speed, which is controlled by the temperature in the active GST, must be

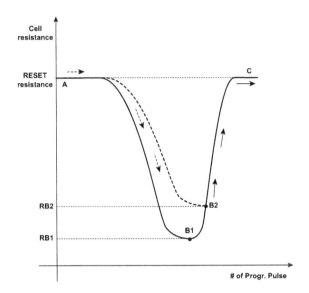

Fig. 8 Evolution of the cell resistance during a SET SCU programming sequence (briefly, SCU programming characteristic). The two plots refer to two programming sequences having the same constant value of δV_{SCU} and different values of $t_{P,SCU}$. *Continuous-line* long pulses; *dashed-line* short pulses

sufficiently high to fully crystallize the GST before the pulses of the SCU reach a critical amplitude that makes the GST start amorphizing. This means that a complex trade-off between δV_{SCU} and $t_{P,SCU}$ exists. If the energy delivered by the SCU programming pulses before they reach the critical amplitude is not enough to make the active GST fully crystallize, the cell resistance does not reach its minimum value, and the SET SCU algorithm fails to program the cell. This can happen if the pulse duration $t_{P,SCU}$ is too short, so that the crystallization speed is not adequate to sustain the growth of the crystal fraction during each SCU step: in practice, in this case, the increase in the programming voltage is too fast with respect to the decrease in the cell resistance at each step. In this respect, consider two SCU programming sequences having the same value of δV_{SCU} but different values of pulse length, t_{P1} and t_{P2} (with $t_{P1} > t_{P2}$). The situation is depicted in Fig. 8, where the programming characteristics corresponding to the two different values of $t_{P,SCU}$ are given (curve A-B_1-C: long pulses; curve A-B_2-C: short pulses). It is apparent that, in the case of curve A-B_2-C, it is impossible to program the cell to resistance values in the range from R_{B1} to R_{B2}, which means that the programming algorithm will fail to converge if the target SET resistance is lower than R_{B2}.

An alternative programming approach that overcomes the above drawback is to implement a partial-RESET SCU programming algorithm. In this case, the cell is initially crystallized (for instance, through a SET SCU pulse) and then partially amorphized by applying a sequence of RESET pulses with increasing amplitude and constant time duration (Fig. 9). In practice, the situation is complementary with respect to the case of partial-SET SCU programming. In fact, crystallization and amorphization act in an opposite manner: in the partial-SET SCU algorithm, the target resistance is obtained by exploiting crystallization, whereas the partial-RESET SCU algorithm exploits amorphization. Partial-SET SCU programming works in the

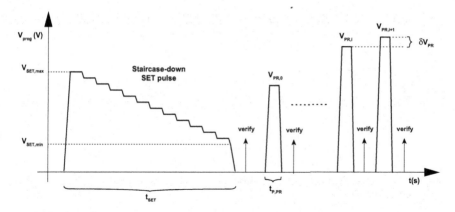

Fig. 9 Time diagram of a partial-RESET SCU programming sequence. The cell is initially crystallized through a SCD SET pulse and, then, the partial amorphization starts controlled by RESET pulses with increasing amplitude. Each programming pulse is followed by a verify operation

portion A → B1 of the programming characteristic in Fig. 8, whereas partial-RESET SCU works in the portion B1 → C of the characteristic. It is worth pointing out that, even though by using the two algorithms it is possible to program the cell to any target resistance value R_T in the allowed range, nonetheless, the phase distribution in the active GST at the end of the two programming sequences will be completely different (see Sect. 4).

The main advantage of partial-RESET SCU programming is that, differently from the case of partial-SET SCU programming, the evolution of the cell resistance evolves in any case from B_1 to C (provided that the initial SCD SET pulse is well defined so as to allow the complete crystallization of the GST), thus enabling a better control of the cell resistance. The negative aspect of this algorithm is the need for the large amount of current required to partially amorphize the GST during each programming pulse (indeed, a higher pulse amplitude must be provided).

In some practical cases, the ML algorithm can be made by a suitable combination of the two above approaches. Indeed, the two solutions have particular advantages and disadvantages when considering the specific value of the resistance level to be programmed.

An interesting aspect in ML PCM programming is the placement strategy for the programmed distributions [21]. The choice of the spacing between adjacent programmed levels within the available window is driven by trade-off considerations between program and read operations. From the viewpoint of programming, the parameter that is directly controlled when applying a partial-RESET pulse to the cell is the thickness of the obtained amorphous cap that, to a first approximation, determines the cell resistance through a direct-proportionality relationship (this assumption holds for relatively small cap thicknesses, as in our case, whereas the resistance has a more complex dependence on the amorphous cap thickness for increasing values of this parameter). Then, the intrinsic mechanism of

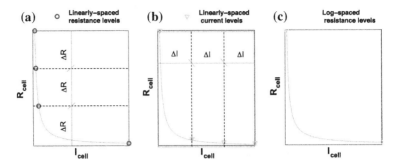

Fig. 10 ML spacing strategies. **a** Equally spaced resistance levels. **b** Equally spaced read current levels. **c** Log-spaced levels

partial-RESET programming suggests choosing equally spaced levels within the available resistance window, as shown in Fig. 10a, where, for simplicity, the case of a 2-bit/cell storage is considered. However, when adopting a current sensing approach, as in our case, this choice results in current levels very close to each other in the lower part of the read current window, which implies long sensing time to achieve the necessary accuracy.

On the other hand, the optimum choice from the sensing viewpoint, that is to equally space the programmed levels within the available read current window (Fig. 10b), makes programming accuracy requirements more severe for low-resistance levels, which are closer to each other, thus resulting in longer programming time.

A reasonable trade-off between programming and readout requirements is to choose log-spaced levels within the available resistance window, as depicted in Fig. 10c. This way, the intermediate read current levels are shifted toward higher and more spaced values when compared to the case of equally-spaced resistance levels, thus relaxing sensing accuracy requirements, and, simultaneously, the minimum required difference in the cell resistance (and, hence, in the amorphous cap thickness) between adjacent levels is higher with respect to the case of equally-spaced, read current levels. Furthermore, using log-spaced levels is beneficial with respect to the case of equally-spaced resistance levels when considering the effect of resistance drift, which is more severe in the case of higher cell resistance.

4 Phase Distribution in ML Programming

As mentioned above, the partial-SET and the partial-RESET SCU programming algorithms lead to different phase distributions in the active GST volume at the end of the programming operation. Let us consider the diagram in Fig. 11, where the evolution of the read cell current is provided as a function of the SCU voltage

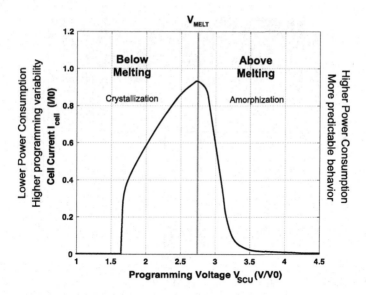

Fig. 11 Programming characteristic of a PCM cell (SCU programming algorithm). V_0 and I_0 are normalizing constants. The cell is initially in the RESET state. The characteristic is divided into two parts. In the part on the *left side*, the cell resistance is controlled by crystallization (decreasing resistance and, hence, increasing read cell current for increasing values of V_{SCU}) whereas, in the part on the *right side*, the resistance evolution is controlled by amorphization (increasing resistance and, hence, decreasing read cell current for increasing values of V_{SCU})

applied to a cell that was initially programmed to the RESET state (the cell current is measured after each SCU programming pulse). To a first-order approximation, the temperature at the heater-GST interface, T_{int}, in the presence of a pulse with amplitude V_{SCU} can be expressed as:

$$T_{int} = \frac{R_{th}}{R_H} V_{SCU}^2 + T_{amb} \qquad (9)$$

where R_{th} and R_H are the thermal resistance of the cell and the electrical resistance of the heater, respectively, and T_{amb} is room temperature. T_{int} is, typically, the maximum temperature inside the memory cell for any value of the applied V_{SCU} (see the simulated temperature profile in Fig. 12) [31]. As long as the programming voltage is lower than the critical value V_M that causes the temperature inside the GST to reach the melting point T_M, crystallization will be the only process responsible for phase transition: the higher the applied pulse amplitude, the higher the temperature inside the GST and, hence, the bigger the obtained crystal fraction and the lower the achieved cell resistance. Once the programming voltage reaches (and overcomes) the critical value V_M, the temperature inside the GST reaches (and overcomes) T_M and the amorphization process will start dominating the evolution of the phase distribution inside the active GST (this operating region corresponds to

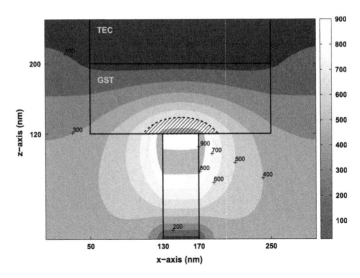

Fig. 12 Simulated temperature profile in a PCM cell during programming (a pulse amplitude V_{SCU} higher than V_M is assumed). The temperature is maximum at the heater-GST interface. The *dashed line* defines the region where the temperature is equal to the melting temperature, T_M. The GST in the region enclosed by the *dashed line* is melted and will be amorphized in case the applied voltage is sharply decreased (fast quenching). The thickness, x_a, of the molten GST and, hence, of the amorphous cap above the heater, is a function of the applied programming pulse: see (10)

the partial-RESET programming domain). The thickness of the amorphized cap above the heater obtained after a fast quenching can be expressed approximately as:

$$x_A = h \frac{(T_{int} - T_M)}{T_M - T_{amb}} \tag{10}$$

where h is the thickness of the overall GST layer.

Since T_{int} is controlled by the amplitude of the programming pulse and is not dominated by random processes, (9) and (10) state that programming the cell resistance by means of the partial-RESET approach gives more predictable results than programming in the region where the cell evolution is dominated by crystallization. Indeed, each increase of the applied voltage will result in a "controlled" increase of the amorphous cap thickness and, hence, of the cell resistance (since the resistivity of the amorphous phase is much larger than the resistivity of the crystalline phase, we can roughly assume that the resistance of the cell is directly proportional to x_A (at least for relatively small cap thicknesses), as mentioned above).

In practice, the evolution of the phase distribution during programming can be summarized as shown in Fig. 13, where two possible alternative phase configurations inside the GST after programming are schematically illustrated [32–34]. The picture on the left side corresponds to a parallel-like phase distribution (where a

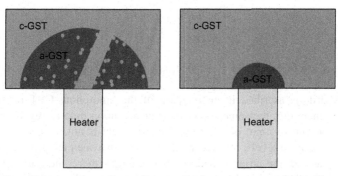

Fig. 13 Conceptual phase distribution inside the GST after programming. Parallel-like phase distribution (*left*): a conductive path is dug and grows inside the amorphous GST; series-like phase distribution (*right*): an amorphous cap grows inside the crystalline GST; a-GST (*red*): amorphous GST; c-GST (*blue*): crystalline GST

conductive path is dug inside the amorphous GST), whereas the picture on the right side represents a series-like phase distribution. By programming the cell with a partial-SET SCU algorithm (left side of the diagram in Fig. 11), the evolution of the phase distribution will be better described by a parallel-like model (the thickness of the amorphous cap where the conductive path will be dug is determined by the initial RESET pulse). In contrast, in the case of partial-RESET SCU programming (right side of the diagram in Fig. 11), the phase distribution can be represented by a series-like model, where the amorphous cap on top of the heater increases for increasing amplitudes of the programming pulse (it is worth to recall that, in partial-RESET SCU programming, the GST is initially fully crystallized by a SCD SET sequence).

4.1 Drift Dependence on Programmed Resistance

The programmed resistance of a PCM cell changes over time due to two physical phenomena: the crystallization process of amorphous GST and the drift of the amorphous-GST resistivity. Since the former leads to a resistance decrease with time, it is the only responsible for data loss in bi-level storage. On the contrary, in multi-level storage, both the above phenomena affect data retention, since they may cause problems in distinguishing two adjacent intermediate resistance levels. In particular, the drift also represents a severe limit for the implementation of P&V algorithms since the verify operation after a programming pulse can be affected by the short-term drift of GST resistance. At the cell level, the resistance drift follows the well-known power-law [35]:

$$\frac{R(t)}{R_0} = \left(\frac{t}{t_0}\right)^{v} \qquad (11)$$

where R_0 is a reference GST resistance, t_0 is a normalizing constant, and v is the drift dynamics exponent (briefly, drift exponent). The main contribution to the resistance drift is ascribed to an increase of the amorphous-GST resistivity ρ_A which has, therefore, the same power dependence upon time as given by (11).

It is important to point out that the drift exponent is not the same for all programmed resistance levels, but a relationship exists between the drift exponent and the actual value of the programmed resistance. Furthermore, the drift also depends on the phase distribution inside the active GST. Namely, when considering a specific resistance R_T obtained with partial-SET or partial-RESET SCU programming algorithm, the measured drift exponent will be different in the two cases. This is ascribed to the fact that, as discussed in previous section, the distribution of the crystalline and the amorphous phase in the active GST follows either a parallel- or a series-like topology depending on the used programming algorithm. Since the drift mainly affects the resistivity of the amorphous phase, having a parallel-like or a series-like phase distribution in the active GST will result in a different drift of the overall cell resistance (namely, the resistance obtained with the partial-RESET approach will exhibit a higher drift exponent due to the presence of the series-like phase distribution).

When considering, for instance, the case of cells programmed by means of a partial-RESET SCU algorithm, the drift exponent ranges from less than 0.01 to about 0.1 [36–38], the minimum value being observed in the case of a GST resistance slightly higher than that of the fully crystalline GST layer. In particular, if we consider two different resistance levels, $R_{GST,1}$ and $R_{GST,2}$, comprised between the fully crystalline and the fully amorphous state, it is possible to write the following relationship:

$$v_1 - v_2 \propto \ln\left(\frac{R_{GST,1}}{R_{GST,2}}\right) \qquad (12)$$

where v_1 and v_2 are the drift exponents of $R_{GST,1}$ and $R_{GST,2}$, respectively. The drift exponent increases logarithmically with increasing values of the cell resistance up to a saturation value (of about 0.1), which is the drift exponent of the fully amorphous state.

4.2   Temperature Dependence on Programmed Resistance

A key issue in ML PCM storage is to safely perform readout of all programmed states over the whole operating temperature range. From the point of view of its electrical behavior over temperature, GST can be modeled as an intrinsic semiconductor [39] and, hence, changes in the electrical resistance of a PCM cell with temperature may become significant, which sets additional constraints on reading operation.

Assuming an exponential Arrhenius-like model, the temperature dependence of the GST resistance R_{GST} can be expressed as:

$$R_{GST}(T) = R_0 \exp\left(\frac{E_{a,R_0}}{k_B}\left(\frac{1}{T} - \frac{1}{T_0}\right)\right) \tag{13}$$

where k_B is Boltzmann constant, T is absolute temperature, T_0 is a reference temperature, $R_0 = R_{GST}(T_0)$ is the cell resistance at $T = T_0$, and $E_{a,R0}$ is the conduction activation energy of a cell programmed to resistance R_0 at $T = T_0$.

From (13), if we read the cell content by applying a suitable voltage V_{rd} across the GST, the ensuing current I_{GST} will have a temperature dependence given by:

$$I_{GST}(T) = \frac{V_{rd}}{R_0} \exp\left(-\frac{E_{a,R_0}}{k_B}\left(\frac{1}{T} - \frac{1}{T_0}\right)\right) \tag{14}$$

It is worth to mention that, from experimental data, there is a substantially logarithmic dependence of $E_{a,R0}$ upon the read cell current I_{rd} at $T = T_0$. Indeed, it is possible to write the simple relationship [40]:

$$E_a(I_{rd}) = E_a(I_{rd,0}) - m \cdot \ln\frac{I_{rd}}{I_{rd,0}} \tag{15}$$

where $I_{rd,0}$ is a normalizing current (for example, the full-SET-state read current) and m is the slope of the curve E_a-I_{rd} in a semilogarithmic plot, which depends on the physical characteristics of the cell (in particular on the GST composition).

5 Write Algorithms for ReRAMs

Resistive Random Access Memory (ReRAM) is the name of a group of non-volatile memory technologies characterized by an electrically reversible, resistive switching between a low (LRS) and high (HRS) resistance state. While the previously described PCM technology also fits in this definition, the name ReRAM is commonly used when the resistive switching is caused by the growth/shrink of a nano-scaled conducting filament inside an insulating dielectric (as opposed to PCM's bulk-phase change mechanism).

Depending on the nature of the ionic species constituting the filament, the ReRAM family is usually divided into Conductive Bridge RAM (CBRAM) and Oxide RAM (OxRAM) [41]. In the CBRAM case [42], the filament is constituted by a chain of diffusing metal cations (typically Cu^+, Ag^+) injected into a solid electrolyte (an oxide or chalcogenide glass) from a soluble electrode. In OxRAM [43],

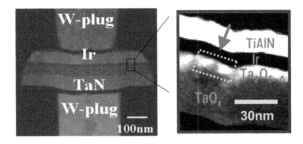

Fig. 14 (*Left*) structure of a ReRAM cell using a Ta_2O_5/TaO_x active stack and TaN and Ir electrodes. (*Right*) TEM image showing the presence of a (substoichiometric) conductive filament into Ta_2O_5 dielectric (Reprinted from [47])

the filament is thought to be constituted by a chain of electrically interactive oxygen vacancies (V_O^+) induced into a transition metal oxide (TMO, typically HfO_2 [44], Ta_2O_5 [45], TiO_2 [46]). This is illustrated in Fig. 14, where a TaOx based OxRAM cell is shown together with the produced Vo^+ filament.

With respect to PCM, the smaller amount of atoms involved in resistive switching, can, in some implementations, lead to faster switching times (below 10 ns) and lower operating voltages (on the order of 1 V) [43]. Additionally, thanks to its filamentary nature, this kind of technology is believed to provide conspicuous advantages in terms of device scaling. ReRAM was therefore initially aggressively investigated as a potential replacement for NAND flash memory. However, due to the recent advances of flash memory regarding vertical 3D integration and the difficulties of controlling ReRAM device behavior at reduced operating current, the application target is gradually migrating towards Storage Class Memory (SCM) or embedded memory applications. In this context, program algorithms are envisaged as a way to minimize program stress on peripheral circuitry or ensure a significant read margin. In Sect. 5.1, focusing on OxRAM devices, we will firstly briefly introduce ReRAM technology and its working principles, while in Sect. 5.3 we will focus more in detail on the main issue of its stochastic variability. Finally in Sect. 5.4, we will review the programming strategies attempted so far.

5.1 ReRAM Technology Overview

Reports of resistance change phenomena in oxide date back to the early 1960s with the pioneering work of Hickmott, Simmons, and Verderber [48, 49]. Although most of the key aspects of this technology (such as the need of forming operation and the presence of a negative differential resistance in the current-voltage characteristic of the memory element) were already outlined in these works, only in the early 2000s had ReRAM started to be aggressively investigated for scaled-semiconductor industrial applications [50]. In general, ReRAM devices are composed of a

Metal-Insulator-Metal (MIM) structure. As fabricated, the structures are thus insulating, and the only possible current through the device is the area-dependent leakage current of the MIM stack. In order to functionalize the material, a first operation called electroforming (or simply forming) needs to be performed. The forming operation consists of a relatively high-voltage (typically >3 V) stress needed to induce a controlled breakdown within the oxide and generate a sufficient amount of O^-, Vo^+ ionic species. After the forming operation, LRS and HRS states can be programmed by a SET and (respectively) a RESET operation using electrical pulses of generally lower voltage and shorter duration.

Depending on the physical mechanism used to RESET the cell, ReRAM technology is often distinguished as unipolar and bipolar. Initial implementations of ReRAM technology were called "unipolar" due to the fact that SET and RESET operation could be operated using the same voltage polarity. The first ReRAM stacks [50], characterized by a symmetric stack geometry and relatively inert metal electrodes (Pt, TiN), often exhibited this behavior. In this case, filament erase (RESET) was obtained, similarly to PCM, by a thermal dissolution of the conductive filament and was thus not linked to a preferential direction of the current flow. This type of operation offers the advantages of unipolar switching operation and a high resistance ON/OFF ratio, but, despite these advantages, unipolar ReRAM gradually lost interest due to the high operating voltage and current, the slower achievable programming speed, and the limited obtainable endurance. Nowadays, modern implementation of ReRAM [44, 45] achieves growth/dissolution of the conductive filament by relying instead on a voltage controlled ion migration. In this approach, SET and RESET operation can performed by fast (down to 10 ns) pulses of approximately similar amplitude but opposite polarity.

5.2 ReRAM Cell Configuration

In the previous description, the ReRAM memory devices have been treated as a two-terminal, "1R" resistor structure. However, such a simple configuration, in itself, cannot be operated reliably using voltage (respectively, current) programming alone. Device current-voltage characteristics during SET operation display an S-shaped negative differential resistance (NDR) [51], and consequently, in voltage controlled programming, require a current-limiting device to prevent device breakdown [52]. Conversely, RESET operation displays an N-shaped NDR and, in current driven programming, may need a voltage limiting device to prevent degradation. Therefore, in real applications, the memory device is usually coupled together with some limiting device (typically a transistor, a diode, or a limiting resistor), thus giving rise to the so-called "1T1R" [44, 52], "1D1R" [53], and "1R1R" [54] cell configurations. A typical cell organization using the 1T1R scheme is depicted in Fig. 15. In this case, the basic 1T1R cell is operated in a voltage-controlled scheme where the (N)MOS selector acts as a (linear) switch during RESET operation and (by controlling saturation current via a properly

Fig. 15 1T1R configuration
during SET (*left*) and RESET
(*right*) operation

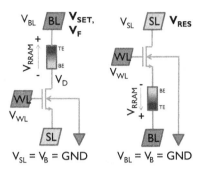

selected wordline—WL—voltage) as a current-limiting device during SET. Therefore, to enable bipolar SET operation, a positive voltage is applied to the selected cell BL while the SL is grounded, and, conversely, to reverse current flow during RESET, a positive voltage is applied to the SL while the BL is grounded. In a well-designed 1T1R structure with controlled stray capacitances, the maximum current reached during a RESET operation is linked to the saturation current imposed by the (N)MOS transistor during SET operation [52] and, for this reason, the latter often generally takes the name of "compliance current" or "operating current". In this configuration, the controlling N(MOS) device needs to fit under the memory element and the effective area of the cell tends to be dictated by the driving-current requirement of the MOSFET rather than by the physical size of the resistive memory element. Nevertheless, it constitutes an effective organization for low-medium density, especially suited for embedded applications.

5.3   The Stochastic Variability Problem in ReRAM

While aggressive scalability and easy manufacturability of filamentary ReRAM has been demonstrated [55], one of the major issues of ReRAM is the stochastic variability in the device operation [56, 57]. In the well-established NAND-flash technology, random device-to-device (D2D) fluctuations of the programmed state occur as a consequence of process related device scaling. In this case, we properly speak of variability of device characteristics such as channel width W, channel length L, drain current I_{DS}, and threshold voltage V_T. ReRAM memories, due to their fundamentally different operating mechanism, additionally display significant intra-device, cycle-to-cycle (C2C) dispersion from the very beginning of device lifetime. In this sense, we should distinguish the typical process variability from the stochastic variability where each characteristic is subject to independent and (ideally) uncorrelated cycle-to-cycle fluctuations. More precisely, the stochastic variability affects both the voltage needed to trigger a SET/RESET transition and the final LRS/HRS programmed state resulting from this transition, and we will therefore speak of resistance variability and voltage variability. While the physical cell size plays a minor role in controlling variability, the operating current is the parameter that has the greatest impact. Unfortunately, operating current reduction worsens device controllability [56, 57].

The resistance distribution for both ON and OFF states can, to a first approximation, be described by means of a log-normal distribution, and, to establish a useful figure of merit, it is possible to consider the standard deviation of resistance distribution σ_R normalized by the median resistance, σ_R/R [56].

Using this criterion, it becomes apparent that, by lowering the operating current, not only the absolute spread increases (as would be expected from a fluctuation proportional to the filament radial cross-section) but also the relative spread increases. In a simplified view, this behavior is consistent with a picture where similar fluctuations in the number of discrete-sized defects constituting the narrowest cross-section of the filament affect a "weaker" resistive filament (where the initial number of defects is low) much more strongly than a "stronger" conductive one (where the change has a lower impact due to larger initial number of defects). In the limit, the filament radial cross-section is so small that the exact spatial arrangement of the conducting path (or in other words its "geometry") affects the conduction much more than the radius at its narrowest cross-section. This is captured by the σ_R/μ_R trend in Fig. 16. When an ion-movement model is considered [58, 59], ions move from pre-existing reservoirs located at TE and BE to enlarge or shrink (and, maybe even, disrupt) the filament. The source of variation in this case is linked to the size and number of discrete defects defining the radius of a single filament. In this sense, the SET action describes the "effort" necessary to "nucleate" a filament either by injecting ions into a gap or by gradually changing the shape of quantum-defined conducting filament.

5.4 ReRAM Program Algorithm, an ISPP Implementation

It is clear from Fig. 16 that, while a window generally exists between the median resistances of the LRS and the HRS state at the array level, the read margin (which takes the resistance distribution tails into account) is unacceptably degraded for operating currents lower than 50 μA. In order to overcome the above mentioned issues, an adaptive programming algorithm would represent a logical choice to ensure enough read margin, while still preserving acceptable power consumption and device stress. The most straightforward implementation of ReRAM program algorithm consists of an adaptation of ISPP (Incremental Step Pulse Programming), already used for flash memory programming [60].

In the ReRAM case, to adaptively program the conductive LRS state during SET, the voltage on the gate of the limiting MOSFET (usually the WL voltage) is stepped from a low value (corresponding to a minimal operating current I_{op} of, e.g., $I_{op} = 20$ μA) to a high value (corresponding to the maximum operating current allowed, e.g., $I_{op} = 100$ μA), while BL is held at constant potential. Conversely, to program the HRS state during RESET, the BL voltage is swept from a rest voltage

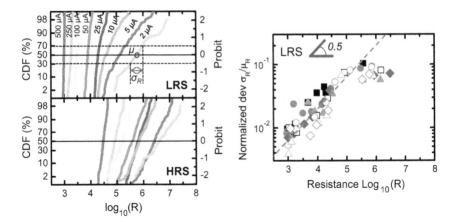

Fig. 16 (*Left*) resistance distribution of LRS and HRS states obtained at various operating currents for a TiN/HfO2/Hf/TiN stack. (*Right*) normalized standard deviation versus median resistance for LRS states (Adapted from [56])

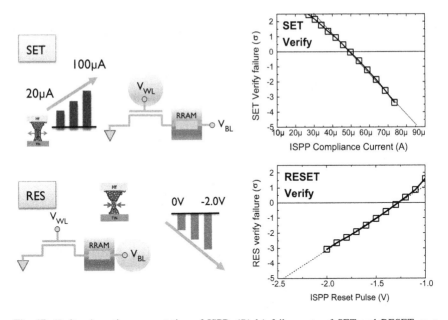

Fig. 17 (*Left*) schematic representation of ISPP. (*Right*) failure rate of SET and RESET as a function of incremental operating current and voltage, respectively. Resistance thresholds of $R_{th, LRS} = 20 \text{ k}\Omega$ and $R_{th,HRS} = 200 \text{ k}\Omega$ are considered for SET and RESET programming, respectively. (Adapted from [60])

(typically 0 V) to a maximum voltage (e.g., -2.0 V). As shown in Fig. 17 [60], this approach is able to initially place the resistance level below (LRS) or above (HRS) a specified resistance threshold.

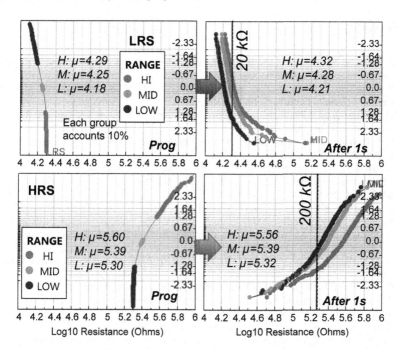

Fig. 18 (*left*) LRS and HRS distribution produced by SET and RESET ISPP algorithm. For each state, three population groups have been highlighted. (*right*) Distribution of the same group of bits after 1 s. (Reprinted from [60])

Unfortunately, a different picture arises if the resistance levels are read 1 s after the last programming step. Figure 18 shows the evolution of LRS and HRS resistances from the verified programmed value to the read-out condition after 1 s. Two observations are apparent. Firstly, the adaptively programmed read margin is completely lost for both the LRS and the HRS state, and, in both cases, distributions basically revert back to their un-verified shape. Secondly, qualitatively no difference in behavior can be traced between the group of bits that underwent an early verify (fast bit) and the bits whose resistance level was placed near the threshold (slow bits). The same phenomena has also been confirmed at the array level for both OxRAM and CBRAM technologies [61]. The above observation suggests that the filament resistance not only stochastically changes within different programming cycles but also spontaneously rearranges in time after the write operation is over. The causes of this behavior have been ascribed to either random telegraph noise (RTN) [61] or spontaneous ion movement [62], however to date no effective program algorithm has been proposed for ReRAM devices.

Acknowledgements The authors wish to thank all the people, the companies, and the institutions, too numerous to list, who have contributed to their research.

References

1. P. Pavan, R. Bez, P. Olivo, and E. Zanoni, "Flash memory cells—An overview," *Proc. IEEE*, vol. 85, no.8, pp. 1248–1271, Aug. 1997.
2. P. Cappelletti, C. Golla, P. Olivo, and E. Zanoni (Eds.), *Flash Memories*. Boston, MA, USA: Kluwer Academic Publishers, 1999.
3. G. W. Burr *et al.*, "Phase change memory technology," *J. Vacuum Sci. Technol. B*, vol. 28, no. 2, pp. 223–262, Mar. 2010.
4. H. S. P. Wong *et al.*, "Phase change memory," in *Proc. IEEE*, vol. 98, no. 12, pp. 2201–2227, Dec. 2010.
5. A. Modelli, A. Manstretta, and G. Torelli, "Basic feasibility constraints for multilevel CHE-programmed Flash memories", *IEEE Trans. Electron Devices*, vol. 48, no. 9, pp. 2032–2042, Sept. 2001.
6. G. Campardo, R. Micheloni, and D. Novosel, *VLSI-Design of Non-Volatile Memories*. Berlin, Germany: Springer-Verlag, 2005.
7. A. Silvagni *et al.* "An overview of logic architectures inside Flash memory devices," *Proc. IEEE*, vol. 91, no. 4, pp. 569–580, Apr. 2003.
8. R. Micheloni, L. Cripa, and A. Marelli, *Inside NAND Flash Memories*. Dordrecht: Springer, 2010.
9. G. Torelli and P. Lupi, "An improved method for programming a word-erasable EEPROM", *Alta Frequenza*, vol. LII, no. 6, Nov./Dec. 1983, pp. 487–494.
10. V. N. Kynett *et al.*, "A 90-ns one-million erase/program cycle 1-Mbit flash memory," *IEEE J. Solid-State Circuits*, vol. 24, pp. 1259–1264, Oct. 1989.
11. T. Tanaka *et al.,* "A quick intelligent page-programming architecture and a shielding bitline sensing method for 3 V-only NAND flash memory," *IEEE J. Solid-State Circuits*, vol. 29, no. 11, pp. 1366–1373, Nov. 1994.
12. R. Shirota *et al.*, "A new programming method and cell architecture for multi-level NAND Flash memories," in *Proc. 14th IEEE Non-Volatile Semiconductor Memory Workshop*, Monterey, CA, USA, Aug. 1995.
13. K. Suh et al., "A 3.3 V 32 Mb NAND flash memory with incremental step pulse programming scheme," *IEEE J. Solid-State Circuits*, vol. 30, no. 11, Nov. 1995, pp 1149–1156.
14. M. Bauer *et al.*, "A multilevel-cell 32 Mb Flash memory," in *1995 IEEE Int. Solid-State Circuits Conf. (ISSCC) Dig. Tech. Papers*, San Francisco, CA, USA, Feb. 1995, pp. 132–133.
15. T.-S. Jung *et al.*, "A 117-mm^2 3.3-V only 128-Mb multilevel NAND flash memory for mass storage applications," *IEEE J. Solid-State Circuits*, vol. 31, no. 11, pp. 1575–1583, Nov 1996.
16. B. Riccó *et al.*, "Nonvolatile multilevel memories for digital applications," *Proc. IEEE*, vol. 86, pp. 2399–2421, Dec. 1998.
17. R. Micheloni *et al.*, "A 0.13-μm CMOS NOR Flash memory experimental chip for 4-b/cell digital storage," in *Proc. 28th European Solid-State Circuits Conf. (ESSCIRC)*, Florence, Italy, Sept. 2002, pp. 131–134.
18. N. Shibata et al., "A 70 nm 16 GB 16-level-cell NAND flash memory," *IEEE J. Solid-State Circuits*, vol. 43, no. 4, pp. 929–937, Apr. 2008.
19. S. Braga, A. Cabrini, and G. Torelli, "Experimental analysis of partial-SET state stability in phase change memories," *IEEE Trans. Electron Devices*, vol. 58, no. 2, pp. 517–522, Feb. 2011.
20. S. Braga, A. Sanasi, A. Cabrini, and G. Torelli, "Modeling of partial-RESET dynamics in Phase Change Memories," in *Proc. 40th European Solid State Device Research Conference (ESSDERC)*, Sevilla, Spain, Sept. 2010, pp. 456–459.
21. S. Braga, A. Sanasi, A. Cabrini, and G. Torelli, "Voltage-driven partial-RESET multilevel programming in phase-change memories," *IEEE Trans. Electron Devices*, vol. 57, no. 10, pp. 2556–2563, Oct. 2010.

22. A. Pirovano *et al.*, "Low-field amorphous state resistance and threshold voltage drift in chalcogenide materials," in *IEEE Trans. Electron Devices*, vol. 51, no. 5, pp. 714–719, May 2004.

23. M. Boniardi *et al.*, "A physics-based model of electrical conduction decrease with time in amorphous $Ge_2Sb_2Te_5$," *J. Appl. Phys.*, vol. 105, no. 8, pp. 084506-1–084506-5, Apr. 2009.

24. N. Papandreou *et al.*, "Drift-tolerant multilevel phase-change memory," in *Proc. 2011 3rd IEEE Int. Memory Workshop (IMW)*, Monterey, CA, USA, May 2011, pp. 1–4.

25. S. Senkader and C. D. Wright, "Models for phase-change of $Ge_2Sb_2Te_5$ in optical and electrical memory devices," *J. Appl. Phys.*, vol. 95, no. 2, pp. 504–511, Jan. 2004.

26. U. Russo, D. Ielmini, A. Redaelli, and A. L. Lacaita, "Intrinsic data retention in nanoscaled phase-change memories—Part I: Monte Carlo model for crystallization and perolation," *IEEE Trans. Electron Devices*, vol. 53, no. 12, pp. 3032–3039, Dec. 2006.

27. F. Bedeschi, C. Boffino, E. Bonizzoni, C. Resta, and G. Torelli, "Staircase-down SET programming approach for phase-change memories", *Microelectronics Journal*, vol. 38, no. 10-11, Oct.-Nov. 2007, pp. 1064–1069.

28. T. Nirschl *et al.*, "Write strategies for 2 and 4-bit multi-level phase-change memory," *2007 IEEE Int. Electron Devices Meeting (IEDM) Tech. Dig.*, Washington, DC, USA, Dec. 2007, pp. 461–464.

29. F. Bedeschi *et al.*, "A bipolar-selected phase-change memory featuring multi-level cell storage," *IEEE J. Solid-State Circuits*, vol. 44, no. 1, pp. 217–227, Jan. 2009.

30. N. Papandreou *et al.*, "Programming algorithms for multilevel phase-change memory," in *Proc. 2011 IEEE Int. Symp. Circuits and Systems (ISCAS)*, Rio de Janeiro, Brazil, May 2011, pp. 329–332.

31. S. Braga, A. Cabrini, and G. Torelli, "An integrated multi-physics approach to the modeling of a phase-change memory device," in *Proc. 38th European Solid-State Device Research Conf. (ESSDERC)*, Edinburgh, UK, Sept. 2008, pp. 154–157.

32. D. Ielmini et al., "Analysis of phase distribution in phase-change nonvolatile memories," *IEEE Electron Device Lett.*, vol. 25, no. 7, pp. 507–509, July 2004.

33. A. Cabrini, S. Braga, and G. Torelli, "Drift-driven investigation of phase distribution in Phase-Change Memories, *Proc. 21st IEEE Int. Conf. Electronics, Circuits, and Systems (ICECS)*, Marseille, France, Dec. 2014, pp. 299–302.

34. N. Papandreou *et al.*, "Estimation of amorphous fraction in multilevel phase change memory cells," in *Proc. 39th European Solid State Device Research Conference (ESSDERC)*, Athens, Greece, Sept. 2009, pp. 209–212.

35. I. V. Karpov *et al.*, "Fundamental drift of parameters in chalcogenide phase change memory," *J. Appl. Phys.*, vol. 102, no. 12, 124503-1–124503-6, Dec. 2007.

36. S. Lavizzari, D. Ielmini, D. Sharma, and A. L. Lacaita, "Structural relaxation in chalcogenide-based phase change memories (PCMs): from defect-annihilation kinetic to device-reliability prediction," in *Proc. Eur. Phase Change Ovonics Symp. (E\PCOS)*, Prague, Czech Republic, 2008.

37. S. Braga, A. Cabrini and G. Torelli, "Dependence of resistance drift on the amorphous cap size in phase change memory arrays," *Appl. Phys. Lett.*, vol. 94, 092112-1–092112-1, March 2009.

38. N. Papandreou *et al.*, "Drift-resilient cell-state metric for multi-level phase-change memory" in *2011 IEEE Int. Electron Devices Meeting (IEDM) Tech. Dig.*, Washington, DC, USA, Dec. 2011, pp. 55–58.

39. A. Redaelli, D. Ielmini, U. Russo, and A. L. Lacaita, "Intrinsic data retention in nanoscaled phase-change memories—Part II: Statistical analysis and prediction of failure time," *IEEE Trans. Electron Devices*, vol. 53, no. 12, pp. 3040–3046, Dec. 2006.

40. A. Cabrini, A. Fantini, F. Gallazzi, and G. Torelli, "Temperature dependence of the programmed states in GST-based multilevel phase-change memories", *Proc. 15th IEEE Int. Conf. Electronics, Circuits and Systems (ICECS)*, St. Julien's, Malta, Aug.-Sept. 2008, pp. 186–189.

41. R. Waser, R. Dittman, G. Staikov, and K. Szot, "Redox-based resistive switching memories – nanoionics mechanism, prospects, and challenges," *Adv. Mater.*, vol. 21, pp. 2632–2663, 2009.

42. I. Valov, R. Waser, J. R. Jameson, and M. N. Kozicki "Electrochemical metallization memories – Fundamentals, applications, prospects," *Nanotechnology,* vol. 22, no. 25, 254003, 2011.

43. H. S. P. Wong *et al.*, "Metal-oxide RRAM," *Proc. IEEE,* vol. 100, no. 6, pp. 1951–2012, Jun. 2012.

44. H. Y. Lee *et al.*, "Low power and high speed bipolar switching with a thin reactive Ti buffer layer in robust HfO$_2$ based RRAM," in *2008 IEEE Int. Electron Devices Meeting (IEDM) Tech. Dig.*, San Francisco, CA, USA, Dec. 2008.

45. Z. Wei *et al.*, "Highly reliable TaOx ReRAM and direct evidence of redox reaction mechanism," in *2008 IEEE Int. Electron Devices Meeting (IEDM) Tech. Dig.*, San Francisco, CA, USA, Dec. 2008, pp. 293–296.

46. D. B. Strukov, G. S. Snider, D. R. Steward, and R. S. Williams, "The missing memristor found," *Nature*, vol. 453, pp. 80–83, May 2008.

47. Z. Wei *et al.*, "Demonstration of high-density ReRAM ensuring 10-year retention at 85°C based on a newly developed reliability model," in *2011 IEEE Int. Electron Devices Meeting (IEDM) Tech. Dig.*, Washington, DC, USA, Dec. 2011, pp. 722–724.

48. T. W. Hickmott, "Low-frequency negative resistance in thin anodic oxide films," *J. Appl. Phys.*, vol. 33, no. 9, pp. 2669–2682, Sep. 1962.

49. J. G. Simmons and R. R. Verderber, "New thin-film resistive memory," *The Radio and Electronic Engineer.*, pp. 81–89, Aug. 1967.

50. I. G. Baek *et al.*, "Highly scalable nonvolatile resistive memory using simple binary oxide driven by asymmetric unipolar voltage pulses," in *2004 IEEE Int. Electron Devices Meeting (IEDM) Tech. Dig.*, San Francisco, CA, USA, Dec. 2004, pp. 587–590.

51. P. Delcroix, S. Blonkowski, and M. Kogelschatz, "Pre-breakdown negative differential resistance in thin oxide film: Conductive-atomic force microscopy observation and modelling," *J. Appl. Phys.*, vol. 110, 034104, 2001.

52. K. Kinoshita *et al.*, "Reduction in the reset current in a resistive random access memory consisting of NiOx brought about by reducing a parasitic capacitance," *App. Phys. Lett.*, vol. 93, 033506, Jul 2008.

53. A. Kawahara *et al.*, "An 8 Mb multi-layered cross-point ReRAM macro with 443 MB/s write throughput," in *Int. Solid-State Circuits Conf. Dig. Tech. Papers*, San Francisco, CA, USA, 2012, pp. 432–434.

54. A. Fantini *et al.*, "Intrinsic switching behavior in HfO$_2$ RRAM by fast electrical measurements on novel 2R test structures," in *Proc. 2012 4th IEEE Int. Memory Workshop (IMW)*, Milano, Italy, May 2011.

55. B. Govoreanu *et al.*, "10 x 10 nm^2 Hf/HfO crossbar resistive RAM with excellent performance, reliability and low-energy operation," in *2011 IEEE Int. Electron Devices Meeting (IEDM) Tech. Dig.*, Washington, DC, USA, Dec. 2011, pp. 729–732.

56. A. Fantini *et al.*, "Intrinsic switching variability in HfO$_2$ RRAM," in *Proc. 2013 5th IEEE Int. Memory Workshop (IMW)*, Monterey, CA, USA, May 2013.

57. S. Ambrogio *et al.*, "Statistical fluctuations in HfOx resistive-switching memory: Part I - set/reset variability," *IEEE Trans. Electron Devices*, vol. 61, no. 8, Aug. 2014, pp. 2912–2919.

58. R. Degraeve *et al.*, "Dynamic 'hour glass' model for SET and RESET in HfO$_2$ RRAM," in *Proc. 2012 Symposium on VLSI Technology*, Honolulu, HI, USA, Jun. 2012, pp. 75–76.

59. S. Larentis *et al.*, "Bipolar-switching model of RRAM by field- and temperature-activated ion migration," in *Proc. 2012 4th IEEE Int. Memory Workshop (IMW)*, Milano, Italy, May 2011.

60. A. Fantini *et al.*, "Intrinsic program instability HfO$_2$ RRAM and consequences on program algorithms," in *2015 IEEE Int. Electron Devices Meeting (IEDM) Tech. Dig.*, Washington, DC, USA, Dec. 2015, pp. 167–172.

61. A. Calderoni *et al.*, "Engineering ReRAM for high-density applications," *Microelectronic Engineering*, vol. 147, Nov. 2015, pp. 145–150.
62. R. Degraeve *et al.*, "Quantitative model for post-program instabilities in Filamentary RRAM," in *Proc. 2016 IEEE Int. Reliability Physics Symposium (IRPS)*, Pasadena, CA, USA, Apr. 2016.

Error Management

Paolo Amato and Marco Sforzin

1 The Role of ECC for Mainstream and Emerging Memory

Over the last four decades memory technologies have evolved and consolidated into two mainstreams: DRAM and NAND flash memories. In DRAMs information is stored by accumulating electrons in capacitors, whereas in NAND flash memories electrons are stored in a floating gate or in an oxide layer in case of *charge trap* NAND. As a consequence, DRAMs have low latency read/write operations, they are volatile and much closer to the CPU in the memory hierarchy. NAND flash memories are high density, non-volatile and more suitable to storage application. Both technologies suffer from the continuous scaling of their electron containers which weakens memory reliability.

On one hand new materials and geometries, such as 3D memories or cross point architectures, are investigated to extend the life of DRAM and NAND. On the other, new memory concepts are emerging [4] where the storage mechanisms are different for each novel type of emerging memory (EM). In Phase Change Memory (PCM) [31] the state is stored in the structure of the material. In metal oxide resistive RAM (Ox-RAM) [18] the state is stored in the oxygen location. In copper resistive RAM (Cu-RAM) [26] it is stored in copper location. In Spin Transfer Torque Magnetic RAM (STTMRAM) [29] it is stored in the electron spin. In Ferroelectric RAM (Fe-RAM) [30] it is stored in the ion displacement. In correlated electron RAM (Ce-RAM or Mott memories) [24] the state is stored in the resistive state of Mott insulators.

P. Amato (✉) · M. Sforzin
Micron Technology Inc, Vimercate, MB, Italy
e-mail: pamato@micron.com

M. Sforzin
e-mail: msforzin@micron.com

© Springer International Publishing AG 2017
R. Gastaldi and G. Campardo (eds.), *In Search of the Next Memory*,
DOI 10.1007/978-3-319-47724-4_8

Fig. 1 the gap between volatile and non-volatile memories may be filled by emerging memories

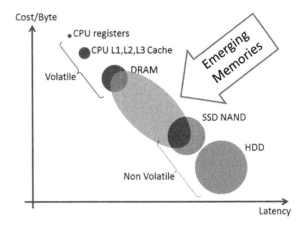

To displace mainstream technologies an EM should show overwhelming advantages. Moreover, EMs based on innovative physical principles are hardly able to reach a high reliability. Yet, emerging technologies can play a fundamental role in improving the memory hierarchy. A classification of consolidated technologies, based on *cost/byte* and performance (*latency*), shows a wide gap between DRAM and NAND (see Fig. 1). The DRAM cost/byte is about 10 times larger than the NAND one, while the NAND latency is 1000 times higher than the DRAM one. The class of storage devices that can fill this gap, referred to as *Storage Class Memories* (SCM) must be both non-volatile/high density and low latency, and offers many opportunities to EMs.

High performance, DRAM-like SCM devices need to be fast and reliable, thus they could benefit from embedded algebraic Error Correction Codes (ECCs) able to correct a few errors in just a few nanoseconds. DRAM is already adopting binary Hamming codes to correct one error per page [12]. For high performance SCM devices it looks natural to extend to two, three, or more bits the protection against errors. For example, in [13] for HfO_X-based resistive memory a bit-error-rate of $\sim 10^{-8}$ is reported. By applying a BCH3 to such bit-error-rate it is possible to achieve DRAM reliability target. As another example, the authors of [7] describe some techniques to improve STTMRAM reliability such that a triple-error-correcting code is enough to achieve the target block failure rate of 10^{-9}.

The main contribution of this chapter is the description of correcting codes able to correct up to two errors, and suitable for being embedded into emerging memory chips.

The two-error correction case has already been treated in [1], and the 3-error correction in [2, 8]. BCH codes are already adopted in NAND flash memories, where they are applied to large pages of thousands of data bits, to correct tens (or even hundreds) of errors. The mild latency constraints of NAND allow traditional BCH decoding. The iterative Berlekamp-Massey (BM) algorithm computes the coefficients of the Error Locator Polynomial (ELP), and the sequential Chien algorithm finds its roots, i.e., the error positions. This classical approach is not compatible with high speed SCM devices: one single iteration of the BM algorithm would require the same latency as the whole decoding process we propose in our design (see [28]). To achieve this challenging latency target, all the iterative and sequential processes of the decoding algorithm must be replaced by full parallel and combinatorial implementations of the same functions. In general, great care is to be taken to avoid time-consuming operations in the Galois Field (GF). A careful optimization of all aspects that can speed up the decoding process is carried on across all stages of the decoding algorithm. The solution shown in this chapter works for any codeword size.

2 Basics of Error Correcting Codes

Channel coding history starts after Shannon proved in 1948 that it is always possible to reliably transmit on a noisy channel provided that the transmission rate is lower than the capacity of the channel (Shannon limit) [16]. That result triggered the research into codes and decoding techniques actually able to reach the Shannon limit. At the beginning two classes of codes were investigated and developed: block codes (decoded by algebraic methods, like Berlekam-Massey and Euclide algorithms) and convolutional codes (decoded by trellis-based methods, like Viterbi and BCJR algorithms). A big step towards the channel capacity has been done by Turbo Codes and Low Density Parity Check Codes (LDPC) [27], which can be considered as belonging to the same class of sparse codes on graphs. Nowadays the research is ongoing into different lines of investigation such as spatially coupled LDPC codes, non binary LDPC codes and polar codes [3].

For our purposes we focus our attention on *linear block codes* [16], such as Hamming and BCH codes, algebraically decoded without resorting to iterative processes.

2.1 Linear Block Codes—Basic Facts

In binary block coding the information is segmented into message blocks of bits of fixed length. Each message block $\mathbf{u} = (u_0, \ldots, u_{k-1})$ consists of k information bits. The encoder, according to certain rules, transforms each binary data vector $\mathbf{u}$ into a binary vector $\mathbf{v}$ of length n (with $n > k$) called *codeword*. The $n - k$ additional bits are called *parity-check bits*.

A code with these characteristics is called a *linear (n, k) code* $\mathscr{C}$ of length n and dimension k. The maximum number of errors that can be corrected by $\mathscr{C}$ is indicated with t. To dig a little more into the correction capability of a linear block codes we introduce a couple of definitions.

The *Hamming weight* $w(\mathbf{x})$ of a vector $\mathbf{x}$ is the number of nonzero components of the codeword.

The *Hamming distance* d_H between two vectors $\mathbf{x}$ and $\mathbf{y}$ is defined as

$$d_H(\mathbf{x}, \mathbf{y}) \triangleq \text{the number of components for which } x_i \neq y_i$$
$$= w(\mathbf{x} - \mathbf{y})$$

Let $\mathscr{C}$ a code of length n. Let's suppose we want $\mathscr{C}$ be able to correct all the errors patterns containing up to t errors (i.e., of weight $\leq t$). If each codeword is sent with the same probability, the best strategy (Maximum Likelihood Decoding) is to pick the codeword "closest" to the received vector $\mathbf{r}$, i.e. the codeword $\mathbf{v}_i$ for which $d_H(\mathbf{v}_i, \mathbf{r})$ is smallest. By using this strategy the code will be capable of correcting all patterns of weight $\leq t$ if and only if the distance between each pair of codewords is $\geq 2t + 1$. In fact if for any couple of codewords $\mathbf{v}_i, \mathbf{v}_j$ we have $d_H(\mathbf{v}_i, \mathbf{v}_j) \geq 2t + 1$ (that is if the Hamming spheres of radius t around $\mathbf{v}_i$ and $\mathbf{v}_j$ are disjoint), then if $\mathbf{v}_i$ is sent and $d_H(\mathbf{v}_i, \mathbf{r}) \leq t$, $\mathbf{r}$ cannot be closer to any other codeword $\mathbf{v}_j$. As example, let's consider a code of length 6. Figure 2 shows two codewords at Hamming distance 3, and their associated Hamming spheres of radius 1.

Thus by defining the *minimum distance of a code* $\mathscr{C}$ as

$$d_{\min}(\mathscr{C}) \triangleq \min\{d_H(\mathbf{v}, \mathbf{v}') : \mathbf{v}, \mathbf{v}' \in \mathscr{C} \wedge \mathbf{v} \neq \mathbf{v}'\},$$

we have that $\mathscr{C}$ is able to correct t errors if and only if $d_{\min}(\mathscr{C}) \geq 2t + 1$.

Figure 2 shows two 6-tuples with Hamming distance 3, and their associated *Hamming spheres*. The figure is a two-dimensional projection of a six-dimensional space.

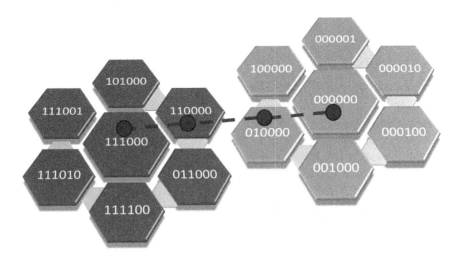

Fig. 2 Two 6-tuples with Hamming distance three, and their associated Hamming spheres of radius one

When it's needed to explicitly identify the correction capability of a linear block code, it's possible to describe it by using the triple (n, k, d_{min}).

2.2 Linear Block Codes Matrix Description

A linear systematic (n, k) block code is completely specified by its $k \times n$ *generator matrix*

$$\mathbf{G} \triangleq [\mathbf{P} \ \mathbf{I}_k],$$

where $\mathbf{I}_k$ denotes the $k \times k$ identity matrix and $\mathbf{P}$ is the $k \times (n - k)$ matrix defining the *parity-check equations*. The function p generating the $n - k$ parity bits can then be defined as

$$p(\mathbf{u}) \triangleq \mathbf{u} \cdot \mathbf{P}$$

Thus the resulting codeword is $\mathbf{v} = [p(\mathbf{u}) \ \mathbf{u}]$.

If the generator matrix $\mathbf{G}$ of $\mathscr{C}$ is in systematic form, the *parity-check matrix* $\mathbf{H}$ may take the following form

$$\mathbf{H} \triangleq [\mathbf{I}_{n-k} \ \mathbf{P}^T]. \tag{1}$$

An n-tuple $\mathbf{v}$ is a codeword of $\mathscr{C}$ if and only if $\mathbf{v} \cdot \mathbf{H}^T = 0$.

Let $\mathbf{v}$ be the codeword written in memory. Due to errors, the n-tuple $\mathbf{r}$ read from memory can be different from $\mathbf{v}$. The vector $\mathbf{e} \triangleq \mathbf{r} + \mathbf{v}$ is called the *error vector*. The aim of any decoding scheme of a block code is to determine the error vector $\mathbf{e}$.

When $\mathbf{r}$ is received, the decoder computes the following $(n - k)$-tuple S

$$S \triangleq \mathbf{r} \cdot \mathbf{H}^T,$$

which is called the *syndrome* of $\mathbf{r}$. We have $S = 0$ if and only if $\mathbf{r}$ is a codeword of $\mathscr{C}$, and $S \neq 0$ if and only if $\mathbf{r}$ is not a codeword of $\mathscr{C}$. Moreover the syndrome S of $\mathbf{r}$ depends only on the error vector $\mathbf{e}$, and not on the original codeword $\mathbf{v}$. In fact

$$S = \mathbf{r} \cdot \mathbf{H}^T = (\mathbf{v} + \mathbf{e}) \cdot \mathbf{H}^T = \mathbf{v} \cdot \mathbf{H}^T + \mathbf{e} \cdot \mathbf{H}^T = \mathbf{e} \cdot \mathbf{H}^T.$$

2.3 Error Correction Performance of Linear Block Codes

The error correction capability t directly impacts the device reliability. In memory jargon, the fraction of bits that contains incorrect data before applying ECC is called the *raw bit error rate* (RBER).

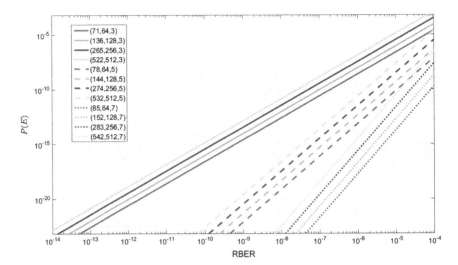

Fig. 3 Log-log plot of $P(E)$ v. RBER for different t and k. Each code is indicated with the triple (n, k, d_{min})

The probability $P(E)$ of *decoder error* (called also *codeword error rate* or *frame error rate*) can be easily bounded by the probability of any error patterns of weight greater than $\lfloor (d_{min} - 1)/2 \rfloor$:

$$P(E) \leq \sum_{i=\lfloor (d_{min}+1)/2 \rfloor}^{n} \binom{n}{i} \text{RBER}^i (1 - \text{RBER})^{n-i}.$$

Figure 3 shows the decoder error as a function of RBER and for different codes. In particular it's worth noticing that $P(E)$ is a weak function of the code length n. To better understand this let's consider a $P(E)$ approximation. If $n\text{RBER} < t$, we have that

$$P(E) \sim \binom{n}{t+1} \text{RBER}^{t+1} \sim \left(\frac{n\text{RBER}}{t+1} \right)^{t+1},$$

and, by applying the logarithm we get

$$\log P(E) \sim (t+1) \log \text{RBER} + (t+1) \log \frac{n}{t+1}.$$

Thus in a log-log plot of $P(E)$ as function of RBER we see a straight line whose slope is $t + 1$. The code length n only influences the line intercept.

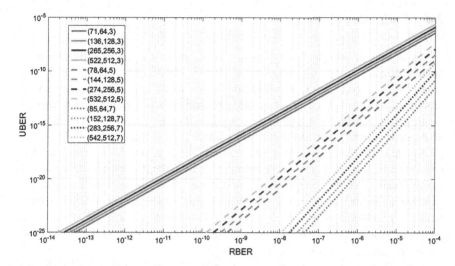

Fig. 4 Log-log plot of Uber v. RBER for different t and k. Each code is indicated with the triple (n, k, d_{min})

A lower bound on the probability of bit error P_b can be obtained by assuming that a decoder error causes a single bit error in the k message bits.[1] On the other hand, an upper bound is obtained by assuming that all k message bits are incorrect when the received vector is incorrectly decoded. Thus

$$\frac{1}{k} P(E) \leq P_b(E) \leq P(E)$$

In memory terminology the error rate after applying ECC is called the *(instantaneous) uncorrectable bit error rate* (UBER). Instantaneous UBER is defined as the probability that a codeword will fail, divided by the number of user bits in the codeword. Thus, by definition, the UBER is a lower bound of the actual bit error probability after decoding.

For instance, Fig. 4 shows UBER as a function of RBER for $t = 1, 2, 3$, for $k = 62, 128, 256$ and 512. A reasonable UBER target for storage applications is 10^{-15}. For such a target an RBER equal to $\sim 6 \cdot 10^{-6}$ can be tolerated with $t = 3$; this RBER is one decade better than $t = 2$, and three decades better than $t = 1$. Vice versa, at RBER of 10^{-8} each additional error the code is able to correct reduces UBER by six decades.

---

[1] $1 - P_b(E)$ is the probability for a single bit to be "good". A codeword is considered "good" if no error is present in the data region, i.e. if all the k information bits are "good". Hence $1 - P(E) = (1 - P_b(E))^k$. For small $P_b(E)$ and large k we have $(1 - P_b(E))^k = 1 - kP_b(E) + o(P_b(E))$ from which directly follows that $P(E) \leq kP_b(E)$.

2.4 Code Modifications

In this section we introduce some minor modifications that can be applied to linear block codes.

An (n, k, d) code is *extended* by adding an additional redundant bit, and so producing an $(n + 1, k + 1, d + 1)$ code. The extension is usually used to increase the detection capability of a code. For example Hamming code are extended to be able to correct one error and detect two errors.

A code is *shortened* by deleting one or more message bits. This means that rows (corresponding to the deleted bits) are removed from the generator matrix, columns (corresponding to the encoded bits) are removed from the generator matrix, and columns (corresponding to the deleted bits) are removed from the parity-check matrix. By shortening $\bar{k}$ bits, an (n, k, d) code becomes a $(n - \bar{k}, k - \bar{k}, d)$ code. The generator matrix $\mathbf{G}$ from $k \times n$ becomes $(k - \bar{k}) \times (n - \bar{k})$, and the parity-check matrix $\mathbf{H}$ from $m \times n$ becomes $m \times (n - \bar{k})$.

As an example, let's take a (7,4) code—this is actually the Hamming (7,4) code that will be described in Sect. 3.2—and suppose we shorten it by removing the first two message bits, obtaining a (5,2) code. The code matrices are modified in the following way:

$$
\mathbf{G} = \begin{bmatrix} 1 & 1 & 0 & 1 & 0 & 0 & 0 \\ 0 & 1 & 1 & 0 & 1 & 0 & 0 \\ 1 & 1 & 1 & 0 & 0 & 1 & 0 \\ 1 & 0 & 1 & 0 & 0 & 0 & 1 \end{bmatrix} \mapsto \mathbf{G} = \begin{bmatrix} 1 & 1 & 1 & 1 & 0 \\ 1 & 0 & 1 & 0 & 1 \end{bmatrix}
$$

$$
\mathbf{H} = \begin{bmatrix} 1 & 0 & 0 & 1 & 0 & 1 & 1 \\ 0 & 1 & 0 & 1 & 1 & 1 & 0 \\ 0 & 0 & 1 & 0 & 1 & 1 & 1 \end{bmatrix} \mapsto \mathbf{H} = \begin{bmatrix} 1 & 0 & 0 & 1 & 1 \\ 0 & 1 & 0 & 1 & 0 \\ 0 & 0 & 1 & 1 & 1 \end{bmatrix}
$$

Almost all the codes used in memory devices are shortened codes, because the message size k is usually a power of two. Especially for full-parallel implementation shortening is an opportunity for optimization. In fact the remaining information bits can be chosen to reduce latency, the number of used gates and sometime also to reduce the number of parity-check bits.

2.5 Technology-Independent Estimates

To compare at abstract level the different ECC solutions, it is convenient to express their decoder areas and latencies in terms of technology-independent units.

This can be obtained by considering only elementary two-input gates as basic building elements, and by assuming the relationship between gates as given in Table 1. These relationships follow from basic reasoning about the internal

Table 1 Technology-independent gate relationship. The basic unit is the NMOS (NCh)

	XOR	NAND	NOR	INV	NCh	PCh
Size	30	8	10	3	1	2
Delay	15	6	6	3	1	1

structure of these elementary gates. In fact, consider the P/N ratio of the CMOS technology equal to 2 (i.e., let $2W/L_{min}$ PMOS be equivalent to W/L_{min} NMOS). Then we may assume an inverter (the NOT gate) has area $2 + 1 = 3$, that is the sum of the areas of the PMOS and the NMOS respectively assuming the area of the NMOS equal to 1, and has delay $2 + 1 = 3$, that is the capacitive load of the output node assuming that each MOS transistor contributes with its own capacitive load proportional to its width. Considering a standard architecture for the other elementary gates, it is possible to assign a size number and a delay number for each of them. Of course this is just an approximation, because for any given elementary gate many different internal architectures can be envisioned, and a specific sizing is set for each of them by taking many parameters into account. Nonetheless, this approach allows to give complexity estimates both in terms of all the different gate types (XOR, NAND, ...), and in terms of a given reference gates. In particular in the following we will use the XOR gate as a reference for the area, and the number of XOR levels as a reference for the delay. The choice of the XOR gate as a reference gate is justified by the fact that it is the most numerous gate in the ECC hardware implementation.

3 Interesting Codes

3.1 Single-Parity-Check Codes

Single-parity-check (SPC) codes [27] are very simple types of linear block codes, able to detect if an odd number of errors has occurred. SPC codes cannot correct errors, but they can be useful in the case the technology has a very low RBER, and the application enables recovery of the data from other sources.

3.1.1 Definition

A SPC $\mathscr{C}_{spc}$ over GF(2) of length n is a binary $(n, n - 1)$ linear code for which each codeword consists of $n - 1$ information bits and a single parity-check bit. Let $\mathbf{u} = (u_0, u_1, \ldots, u_{n-2})$ be the message to be encoded. The encoding consists in adding a single parity-check bit c to form an n-bit codeword $(c, u_0, u_1, \ldots, u_{n-2})$. This single parity-check bit c is simply the XORing (modulo-2 sum) of the $n - 1$ information bits of the message $\mathbf{u}$:

$$c = u_0 + u_1 + \cdots + u_{n-2}$$

The $n \times n$ generator matrix $\mathbf{G}$ of a SPC code in systematic form is then

$$\mathbf{G} = \begin{bmatrix} \begin{matrix} 1 \\ 1 \\ \vdots \\ 1 \\ 1 \end{matrix} & \mathbf{I}_{n-1} \end{bmatrix},$$

while the $1 \times n$ parity-check matrix $\mathbf{H}$ is

$$\mathbf{H} = [\underbrace{1\ 1 \ldots 1\ 1}_{n \text{ times}}].$$

Thus $\mathscr{C}_{\mathrm{spc}}$ consists of all the n-tuples over GF(2) with even weight and hence its minimum distance is 2.

A SPC is not able to correct errors, but it is capable of detecting any error pattern containing an odd number of errors. In fact any such error pattern will change a codeword in $\mathscr{C}_{\mathrm{spc}}$ into a non-codeword. Any error pattern with a nonzero even number of errors, instead, will change a codeword in $\mathscr{C}_{\mathrm{spc}}$ into another codeword.

The decoding process consists of XORing all the bits of the received codeword $\mathbf{r}$ (including the parity check bit). If the result is 0, no error or an even number of errors occurred.

3.1.2 Implementation

Assuming that all the bits of $\mathbf{u}$ are available at the same time, the fastest way for calculating the parity check bit c is to use a XOR tree with $n - 1$ inputs as shown in Fig. 5. Such tree will consists of $n - 2$ XORs, and its depth will be $\lceil \log_2(n - 1) \rceil$ XOR levels. As side note, to re-use the same circuitry also for the decoding process it is simply needed to use a XOR tree with n inputs instead of $n - 1$.

3.2 Hamming Codes

Hamming codes were introduced in 1950 [10]. They are not only excellent prototypes of block codes (they contains most of the properties of most practical codes), but they are also still widely used. For example recently they were integrated into LPDDR4 memory devices [25].

Fig. 5 Xor tree implementing a bit-parallel calculation of the parity check bit c

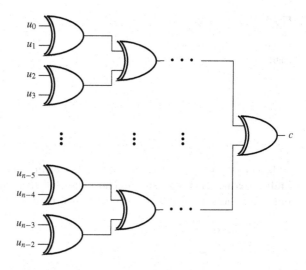

3.2.1 Definition

Let's consider a (n, k) Hamming code with $n = 2^m - 1$. Then $k = n - m = 2^m - m - 1$, i.e. there are m parity-check bits. Hamming codes are able to correct one error, and their minimum Hamming distance is 3.

The parity check matrix of a Hamming code consists of all the nonzero m-tuple (that are exactly 2^{m-1}) as its columns. **H** is composed by $\binom{m}{1}$ columns of weight 1, $\binom{m}{2}$ columns of weight 2, ..., $\binom{m}{m}$ columns of weight m (the m-tuple of all ones). As example, let's take $m = 3$, and then a (7,4) Hamming code. For this code the 3×7 parity check matrix **H** in systematic form is

$$\mathbf{H} = \begin{bmatrix} 1 & 0 & 0 & 1 & 0 & 1 & 1 \\ 0 & 1 & 0 & 1 & 1 & 1 & 0 \\ 0 & 0 & 1 & 0 & 1 & 1 & 1 \end{bmatrix} = \begin{bmatrix} & & 1 & 0 & 1 & 1 \\ \mathbf{I}_3 & & 1 & 1 & 1 & 0 \\ & & 0 & 1 & 1 & 1 \end{bmatrix} = \begin{bmatrix} \mathbf{I}_3 & \mathbf{P}^T \end{bmatrix}.$$

The related 4×7 generator matrix **G** is then:

$$\mathbf{G} = \begin{bmatrix} 1 & 1 & 0 & 1 & 0 & 0 & 0 \\ 0 & 1 & 1 & 0 & 1 & 0 & 0 \\ 1 & 1 & 1 & 0 & 0 & 1 & 0 \\ 1 & 0 & 1 & 0 & 0 & 0 & 1 \end{bmatrix} = \begin{bmatrix} 1 & 1 & 0 & \\ 0 & 1 & 1 & \\ 1 & 1 & 1 & \mathbf{I}_4 \\ 1 & 0 & 1 & \end{bmatrix} = \begin{bmatrix} \mathbf{P} & \mathbf{I}_4 \end{bmatrix}.$$

R. McEliece introduced a simple description of the Hamming code based on Venn diagrams. Figure 6 shows Venn-diagram representation of (7,4) Hamming

Fig. 6 Venn-diagram
representation of (7,4)
Hamming code parity check
matrix

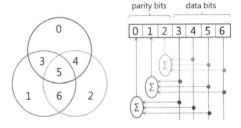

code and a picture that serves to explain its meaning. The encoding rule is translated
in these terms: the parity bits are chosen so that each circle has an even number of
ones, i.e., the sum of bits in each circle is 0 modulo 2.

3.2.2 Decoding

The decoding of a received vector $\mathbf{r}$ can be carried on in three steps [16]:

1. Compute the syndrome S starting form the received vector $\mathbf{r}$ by means of the
   relation $S = \mathbf{r} \cdot \mathbf{H}^T$.
2. Locate the column $\mathbf{h}_i$ of $\mathbf{H}$ which is equal to S. Then the n-tuple $\mathbf{e}$ containing a
   single 1 in the i-th position is assumed to be the error pattern caused by the
   channel.
3. Decode the received vector $\mathbf{r}$ into $\hat{\mathbf{v}} = \mathbf{r} + \mathbf{e}$.

The described decoding procedure, shown also in Fig. 7, is a particular instance
of *syndrome decoding* (called also *table-lookup decoding*).

Fig. 7 General table-lookup
decoder workflow for a linear
block code

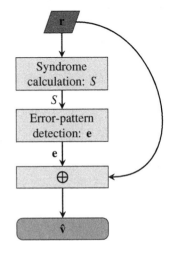

3.2.3 Fully Parallel Implementation

Table-lookup decoding can be implemented with a full parallel approach (by a combinatorial logic circuit).

In a completely bit-parallel implementation each bit of the syndrome $S = (s_0, \ldots, s_{m-1})$ of the received vector $\mathbf{r}$ can be obtained by a separate XOR tree (like for SPC codes) with input taken from the bits of $\mathbf{r}$

$$s_j = \sum_{i=0}^{n-2} r_i h_{ji}$$

being $\mathbf{H} = (h_{ij})$ with $i = 0, \ldots, m-1$ row index and $j = 0, \ldots, n-2$ column index. The effective number of XOR-ed input to generate s_j is the weight w_i of the $n-1$-tupla $col_i(\mathbf{H}^T) = row_i(\mathbf{H}) = h_{ji}|_{j=0,\ldots,n-2}$. The number of elementary XOR2 gates composing a big XOR with w_i inputs is w_i-1 so that the area of the syndrome logic in terms of XOR2 gates is:

$$\sum_{i=1}^{m-1} (w_i - 1) = \left(\sum_{i=1}^{m-1} w_i\right) - m = w(\mathbf{H}) - m$$

where $w(\mathbf{H})$ is the weight of the binary matrix $\mathbf{H}$. This weight is easy to count because it correspond to the total weight of all the binary m-tuple (the zero m-tuple has weight 0 and do not make any contribution) that is given by $m2^m/2 = m2^{m-1}$ (there is the same number of zeros and ones). Consequently the total number of XORs needed for the syndrome calculation is

$$m2^{m-1} - m = m(2^{m-1} - 1).$$

The depth of the logic in terms of elementary XOR 2 to go through is instead the logarithm of the number of inputs of the big XORs, i.e.

$$\lceil \log_2 2^{m-1} \rceil = m - 1.$$

In case of shortened codes, this area and delay can be minimized by taking as columns the m-tuples with the smallest possible weight.

Once the syndrome is evaluated, each bit i of the error pattern $\mathbf{e}$ can be obtained by: (i) performing the bitwise XOR (modulo-2 sum) between the bits of the syndrome S and those of $\mathbf{h}_i = (h_{i,0}, \ldots, h_{i,m-1})$, the i-th column of the parity check matrix $\mathbf{H}$; and (ii) checking that the result is the m-tuple $(1, 1, \ldots, 1, 1)$, by using an m-input AND gate. Algebraically we may represent this operation by the expression:

$$e_i = \prod_{j=0}^{m-1} (s_j h_{i,j} + \bar{s}_j \bar{h}_{i,j}).$$

In fact the HW implementation will not follow blindly the previous expression. It is more convenient to invert the s_j in case of $h_{i,j} = 0$ (or not, in case $h_{i,j} = 1$) than evaluating the expression $(s_j h_{i,j} + \bar{s}_j \bar{h}_{i,j})$ every time, employing for this reason a physical gate. Then, apart from these m common inverters of s_j, we have an m-input AND tree for each column of **H**. The depth of these trees is $\lceil \log_2 m \rceil$ AND 2 levels, and the total number of needed AND2 gates is $k(m-1) = (2^m - m - 1)(m - 1)$.

These area results are not optimized because the intermediate terms in each m - AND realized by AND2 elementary gates could be reused. For example, we should reuse the first stage of the comparators. For every couple of inputs ($\lfloor m/2 \rfloor$ couples) we use 4 AND2 ($AB, A\bar{B}, \bar{A}B, \bar{A}\bar{B}$), for a total of $4\lfloor m/2 \rfloor$ AND2. These gates replace the first stage of all the comparators ($k\lfloor m/2 \rfloor$ gates). The total area in terms of AND2 then becomes:

$$(m - 1 - \lfloor m/2 \rfloor)k + 4\lfloor m/2 \rfloor$$

that we can write explicitly for the primitive version of the code as:

$$(m - 1 - \lfloor m/2 \rfloor)(2^m - m - 1) + 4\lfloor m/2 \rfloor$$

To reduce area occupancy and latency, the same logical function of the AND trees can be implemented by using trees of NAND and NOR gates.

Finally the correction step is achieved by using k XOR gates, with a delay of 1 XOR gate. Table 2 summarize the cost of implementing a primitive Hamming decoder as a function of m in terms of sums (XOR) and products (AND) in $GF(2)$. In Table 3 we consider also the case of the shortened Hamming to the nearest power of two for data.

Table 2 Area and delay for primitive Hamming codes

Block	Area	Latency
Synd	$2m(2^{m-2} - 1) + m$XOR	$(m-1)$XOR
Recognition	$(m - 1 - \lfloor m/2 \rfloor)(2^m - m - 1) + 4\lfloor m/2 \rfloor$AND	$\lceil \log_2 m \rceil$AND AND
Correction	$(2^m - m - 1)$XOR	1XOR

Table 3 Area and delay for shortened Hamming codes

Block	Area	Latency
Synd	$\sum_{i=2}^{\lceil m/2 \rceil - 1} i \binom{m}{i} + \lceil m/2 \rceil \left(m + 1 + \dfrac{\binom{m}{m/2}}{2} [m\ even] \right)$ XOR	$\lceil \log_2 \dfrac{\sum_{i=2}^{\lceil m/2 \rceil} i \binom{m}{i}}{m} + 1 \rceil$XOR
Recognition	$(m - 1 - \lfloor m/2 \rfloor)2^{m-1} + 4\lfloor m/2 \rfloor$AND	$\lceil \log_2 m \rceil$AND
Correction	2^{m-1}XOR	1 XOR

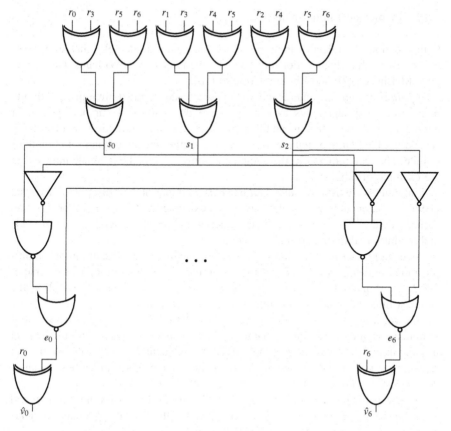

Fig. 8 A full parallel implementation of the decoder of the (7,4) Hamming code

The critical path for a primitive hamming code decoder expressed in terms of XOR and AND to go trough is

$$\lceil log_2 m \rceil \text{ AND} + m \text{ XOR}$$

Figure 8 shows the general full parallel circuit for the decoder of the (7,4) Hamming code.

3.3 BCH Codes

BCH (Bose Chaudhuri Hocquenghem) codes form a large class of cyclic codes for correcting random errors. BCH codes are specified in term of the roots of a polynomial (called *generator polynomial*) over a given *finite field*. Thus, before defining BCH, we briefly review key notions about finite fields.

3.3.1 Primer on Finite Fields

In this section we collect, without proof, the facts about finite fields that are needed in the rest of this chapter. For more details, we refer to MacWilliams and Sloane [19], McEliece [22], and Lin and Costello [16].

A *field* $\mathbb{F}$ is a set of elements in which you can add, subtract, multiply and divide in a way that is analogous to what you can do in the real number system. In particular, each field has an additive identity 0, and a multiplicative identity 1. A field is called *finite* or *infinite* according to whether the underlying set is finite or infinite. The real numbers $\mathbb{R}$, the complex numbers $\mathbb{C}$, and the rationale numbers $\mathbb{Q}$ are example of infinite fields.

Finite fields are usually called *Galois fields*, in honor of their discoverer, Évariste Galois. A Galois field with q elements may be denoted as $\mathrm{GF}(q)$. Let p be a prime, then the field $\mathrm{GF}(p)$ is obtained by considering the set of integer $\{0, 1, \ldots, p-1\}$, and modulo-p addition and multiplication.

As example, let $p = 2$. Then the set $\{0,1\}$ under modulo-2 addition and multiplication as given in Table 4 forms the field $\mathrm{GF}(2)$. Note that the additive inverse of $\mathrm{GF}(2)$ are 0 and 1 themselves. This means that $1 - 1 = 1 + 1 = 0$. $\mathrm{GF}(2)$ is the simplest finite field, and it is commonly referred to as *binary field*.

It's worth noting that $\mathbb{Z}_4$, the set of integers $\{0,1,2,3\}$ equipped with modulo-4 arithmetic, is not a field. This is due to the fact that in a field each non-null element has a multiplicative inverse. But multiplying the element $2 \in \mathbb{Z}_4$ by any element in $\mathbb{Z}_4$, we get just 0 and 2; thus 2 does not have an inverse. In particular, since $2 \cdot 2 = 0$, 2 is said to be a *zero divisor*. Actually Z_4 is a *ring*.

However it is possible to construct the field $\mathrm{GF}(4)$. In fact for any prime p and any positive integer m, there exist the Galois field $\mathrm{GF}(p^m)$ with p^m elements. Thus by taking $p = 2$ and $m = 2$ we get $\mathrm{GF}(2^2)$.

The elements of $\mathrm{GF}(q^m)$, extension of $\mathrm{GF}(q)$, can be represented as m-tuple of elements of $\mathrm{GF}(q)$

$$\mathbf{a} = (a_0, a_1, \ldots, a_{m-2}, a_{m-1}),$$

or as polynomials of degree $m - 1$ with coefficients in $\mathrm{GF}(q)$

$$\mathbf{a}(x) = a_0 + a_1 x + \cdots + a_{m-2} x^{m-2} + a_{m-1} x^{m-1}.$$

Addition in $\mathrm{GF}(q^m)$ is performed element-by-element by using the rules of $\mathrm{GF}(q)$. This means that in $\mathrm{GF}(2^m)$ we have $\mathbf{a} + \mathbf{a} = \mathbf{0}$.

Table 4 Modulo-2 addition and multiplication

+	0	1	·	0	1
0	0	1	0	0	0
1	1	0	1	0	1

Multiplication requires the introduction of a polynomial $\pi(x)$, called *primitive polynomial*, of degree m and *irreducible*; i.e., it is not the product of two polynomials of lower degree with coefficients from $GF(q)$. Irreducible polynomials play, in ring of polynomials, the same role played by prime numbers in the ring of integers.

The product $\mathbf{c} = \mathbf{a} \cdot \mathbf{b}$ with $\mathbf{a}, \mathbf{b}, \mathbf{c} \in GF(q^m)$ is obtained by multiplying the polynomials $\mathbf{a}(x)$ and $\mathbf{b}(x)$, and by taking the remainder modulo $\pi(x)$. Since the degree of $\pi(x)$ is m, the degree of the remainder is at most $m - 1$.

$$\mathbf{c}(x) \equiv \mathbf{a}(x)\mathbf{b}(x) \pmod{\pi(x)}$$

An element $\alpha \in GF(2^m)$ is said to be a *primitive element* of $GF(2^m)$ if the smallest integer i for which $\alpha^i = 1$ is $2^m - 1$. In other words, α generates all the element of the field. The primitive element enables a third representation of the field, called *power representation*, in which each element of $GF(2^m)$ is represented as a power of α. This representation makes multiplication particularly easy. The conversion from power to polynomial representation is obtained by computing $\alpha^i \bmod \pi(\alpha)$.

A fourth representation, used only to represent each m-tuple in a compact way, is the *integer representation*. The integer representation of $\mathbf{a} = (a_0, a_1, \ldots, a_{m-2}, a_{m-1}) \in GF(p^m)$ is simply the integer $a_0 + a_1 p + \cdots + a_{m-2} p^{m-2} + a_{m-1} p^{m-1}$.

As examples, Table 5 shows the finite field $GF(2^2)$, Table 6 shows the finite field $GF(2^4)$, and Table 7 shows the finite field $GF(2^5)$. The tables show the power representation (left), the integer representation (center), and the m-tuple and polynomial representation (right).

The *characteristic* of a field is the smallest integer c such that $c = 1 + 1 + \cdots + 1(c \text{ times}) = 0$. The characteristic of the field $GF(p^m)$ is p. An interesting property of a field of characteristic p, called "freshman's dream" by van Lint [17], is the equation

$$(\mathbf{a} + \mathbf{b})^p = \mathbf{a}^p + \mathbf{b}^p, \tag{2}$$

valid for any element $\mathbf{a}, \mathbf{b} \in GF(p^m)$. By recursively applying it we also have, for each integer k

$$(\mathbf{a} + \mathbf{b})^{p^k} = \mathbf{a}^{p^k} + \mathbf{b}^{p^k}$$

Table 5 $GF(2^2)$ generated by the primitive polynomial $\pi(x) = 1 + x + x^2$ over $GF(2)$

Powerrepresentation	Integerrepresentation	1	α
0	0	0	0
1	1	1	0
α	2	0	1
α^2	3	1	1
$\alpha^3 = 1$			

Table 6 $GF(2^4)$ generated by the primitive polynomial $\pi(x) = 1+x+x^4$ over $GF(2)$

Powerrepresentation	Integerrepresentation	1	α	α^2	α^3
0	0	0	0	0	0
1	1	1	0	0	0
α	2	0	1	0	0
α^2	4	0	0	1	0
α^3	8	0	0	0	1
α^4	3	1	1	0	0
α^5	6	0	1	1	0
α^6	12	0	0	1	1
α^7	11	1	1	0	1
α^8	5	1	0	1	0
α^9	10	0	1	0	1
α^{10}	7	1	1	1	0
α^{11}	14	0	1	1	1
α^{12}	15	1	1	1	1
α^{13}	13	1	0	1	1
α^{14}	9	1	0	0	1
$\alpha^{15} = 1$					

3.3.2 Definition

The construction of a t-error-correcting BCH code (in short BCHt) begin with a Galois Field $GF(2^m)$. Let α be a *primitive element* in $GF(2^m)$. The generator polynomial $\mathbf{g}(x)$ of the t-error-correcting binary BCH code of length $n = 2^m - 1$ is the minimum-degree polynomial over $GF(2^m)$ that has as roots $2t$ consecutive powers of α. It can be shown [16] that the degree of $\mathbf{g}(x)$ is at most mt. This means that the number of parity-check bits of the code is at most mt. There is no simple formula for exact enumeration of the number of parity-check bits. Hovever if n is large and t is small, the number of parity-check bits is exactly mt [20]. Moreover even for larger value of t, the quantity mt is an upper bound and can be assumed as a good approximation of the actual parity-check bits number.

Let's consider the polynomial representation $\mathbf{u}(x)$ of the message $\mathbf{u}$. The code polynomial $\mathbf{v}(x)$ is obtained by multiplying $\mathbf{u}(x)$ by the generator polynomial $\mathbf{g}(x)$. This polynomial multiplication does not lead to systematic encoding, but the encoding can be made systematic by a few other steps that we do not describe here.

When the first of the $2t$ consecutive roots is α, the BCH code is said to be *narrow sense*. If $n = 2^m - 1$ then the BCH code is said to be *primitive*. In the following, we will focus on binary narrow-sense primitive BCH codes.

Table 7 GF(2^5) generated by the primitive polynomial $\pi(x) = 1 + x^2 + x^5$ over GF(2)

Powerrepresentation	Integerrepresentation	1	α	α^2	α^3	α^4
0	0	0	0	0	0	0
1	1	1	0	0	0	0
α	2	0	1	0	0	0
α^2	4	0	0	1	0	0
α^3	8	0	0	0	1	0
α^4	16	0	0	0	0	1
α^5	5	1	0	1	0	0
α^6	10	0	1	0	1	0
α^7	20	0	0	1	0	1
α^8	13	1	0	1	1	0
α^9	26	0	1	0	1	1
α^{10}	17	1	0	0	0	1
α^{11}	7	1	1	1	0	0
α^{12}	14	0	1	1	1	0
α^{13}	28	0	0	1	1	1
α^{14}	29	1	0	1	1	1
α^{15}	31	1	1	1	1	1
α^{16}	27	1	1	0	1	1
α^{17}	19	1	1	0	0	1
α^{18}	3	1	1	0	0	0
α^{19}	6	0	1	1	0	0
α^{20}	12	0	0	1	1	0
α^{21}	24	0	0	0	1	1
α^{22}	21	1	0	1	0	1
α^{23}	15	1	1	1	1	0
α^{24}	30	0	1	1	1	1
α^{25}	25	1	0	0	1	1
α^{26}	23	1	1	1	0	1
α^{27}	11	1	1	0	1	0
α^{28}	22	0	1	1	0	1
α^{29}	9	1	0	0	1	0
α^{30}	18	0	1	0	0	1
$\alpha^{31} = 1$						

As an example let's consider $m = 4$, then $n = 15 = 2^4 - 1$. For $t = 2$ (i.e. a double-error correcting code) the generator polynomial of the code is $\mathbf{g}(x) = 1 + x^4 + x^6 + x^7 + x^8$. Since the number of parity-check bits is equal to the degree of $\mathbf{g}(c)$, we have that $n - k = 8 = mt$. Consequently we get a (15,7) BCH2 code.

For $t = 3$ the generator polynomial is $\mathbf{g}(x) = 1 + x + x^2 + x^4 + x^5 + x^8 + x^{10}$. Thus $n - k = 10 < mt = 12$. The resulting code is a (15,5) BCH3.

Consider a narrow-sense primitive t-error-correcting BCH code $\mathscr{C}$ constructed using GF(2^m), whose generator polynomial $\mathbf{g}(x)$ has $\alpha, \alpha^2, \ldots, \alpha^{2t}$ and their conjugates[2] as roots. Then for $1 \leq i \leq 2t$, $\mathbf{g}(\alpha^i) = 0$. Let $\mathbf{v}(x) = v_0 + v_1 x + \cdots + v_{n-1} x^{n-1}$ be a code polynomial in $\mathscr{C}$. Since $\mathbf{v}(x)$ is divisible by $\mathbf{g}(x)$, a root of $\mathbf{g}(x)$ is also a root of $\mathbf{v}(x)$. Hence, for $1 \leq i \leq 2t$,

$$v(\alpha^i) = v_0 + v_1 \alpha^i + \cdots + v_{n-1} \alpha^{(n-1)i} = 0$$

The above equality can be rewritten in matrix form:

$$(\mathbf{v}_0, \mathbf{v}_1, \ldots, \mathbf{v}_{n-2}, \mathbf{v}_{n-1}) \cdot \begin{bmatrix} 1 & \alpha & \ldots & \alpha^{n-2} & \alpha^{n-1} \\ 1 & \alpha^2 & \ldots & \alpha^{2(n-2)} & \alpha^{2(n-1)} \\ \vdots & \vdots & \ddots & \vdots & \vdots \\ 1 & \alpha^{2t} & \ldots & \alpha^{2t(n-2)} & \alpha^{2t(n-1)} \end{bmatrix}^T = 0. \qquad (3)$$

From (2) for a polynomial $\mathbf{f}(x)$ over GF(2) it holds [16] that

$$\mathbf{f}(\beta^2) = \mathbf{f}^2(\beta).$$

Then we have that $\mathbf{v}(\alpha^{2i}) = \mathbf{v}^2(\alpha^i)$. This means that only the t quantities $\mathbf{v}(\alpha)$, $\mathbf{v}(\alpha^3)$, $\mathbf{v}(\alpha^5), \ldots, \mathbf{v}(\alpha^{2t-1})$ have to be actually evaluated. Thus the matrix in (3) can be substituted by

$$\mathbf{M} \triangleq \begin{bmatrix} 1 & \alpha & \ldots & \alpha^{n-2} & \alpha^{n-1} \\ 1 & \alpha^3 & \ldots & \alpha^{3(n-2)} & \alpha^{3(n-1)} \\ \vdots & \vdots & \ddots & \vdots & \vdots \\ 1 & \alpha^{2t-1} & \ldots & \alpha^{(2t-1)(n-2)} & \alpha^{(2t-1)(n-1)} \end{bmatrix} \qquad (4)$$

The matrix in (4) is the parity check matrix of the t-error-correcting binary BCH code. Since its form is different from that of the parity check matrices introduced in (1), we refer to it with $\mathbf{M}^3$ From (3) to (4) it follows that for any codeword $\mathbf{v}$

$$\mathbf{v} \cdot \mathbf{M}^T = 0 \qquad (5)$$

---

[2] Two elements are conjugates if roots of the same set of polynomials.

[3] Also Chien [6] uses the letter $\mathbf{M}$ to denote such matrix.

Let's consider again the binary (15,7) BCH2. The parity-check matrix $\mathbf{M}$ of this code is

$$\mathbf{M} = \begin{bmatrix} 1 & \alpha & \alpha^2 & \alpha^3 & \alpha^4 & \alpha^5 & \alpha^6 & \alpha^7 & \alpha^8 & \alpha^9 & \alpha^{10} & \alpha^{11} & \alpha^{12} & \alpha^{13} & \alpha^{14} \\ 1 & \alpha^3 & \alpha^6 & \alpha^9 & \alpha^{12} & 1 & \alpha^3 & \alpha^6 & \alpha^9 & \alpha^{12} & 1 & \alpha^3 & \alpha^6 & \alpha^9 & \alpha^{12} \end{bmatrix},$$

where each element of the second row is the third power of the corresponding element of the first row, and we used the fact that $\alpha^{15} = 1$. By using the Table 6, we can write the above matrix in binary format:

$$\mathbf{M} = \left[\begin{array}{ccccccccccccccc} 1 & 0 & 0 & 0 & 1 & 0 & 0 & 1 & 1 & 0 & 1 & 0 & 1 & 1 & 1 \\ 0 & 1 & 0 & 0 & 1 & 1 & 0 & 1 & 0 & 1 & 1 & 1 & 1 & 0 & 0 \\ 0 & 0 & 1 & 0 & 0 & 1 & 1 & 0 & 1 & 0 & 1 & 1 & 1 & 1 & 0 \\ 0 & 0 & 0 & 1 & 0 & 0 & 1 & 1 & 0 & 1 & 0 & 1 & 1 & 1 & 1 \\ - & - & - & - & - & - & - & - & - & - & - & - & - & - & - \\ 1 & 0 & 0 & 0 & 1 & 1 & 0 & 0 & 0 & 1 & 1 & 0 & 0 & 0 & 1 \\ 0 & 0 & 0 & 1 & 1 & 0 & 0 & 0 & 1 & 1 & 0 & 0 & 0 & 1 & 1 \\ 0 & 0 & 1 & 0 & 1 & 0 & 0 & 1 & 0 & 1 & 0 & 0 & 1 & 0 & 1 \\ 0 & 1 & 1 & 1 & 1 & 0 & 1 & 1 & 1 & 1 & 0 & 1 & 1 & 1 & 1 \end{array} \right],$$

From the above equation it's clear that a parity check-matrix $\mathbf{M}$ can be divided into t sub-parity check matrices:

$$\mathbf{M} = \begin{bmatrix} \mathbf{M}_1 \\ \mathbf{M}_3 \\ \mathbf{M}_5 \\ \vdots \\ \mathbf{M}_{2t-1} \end{bmatrix}$$

Two fields are involved in the construction of the BCH codes. The coefficients of generator polynomial and the elements of the codewords are in the "small" field $GF(2)$. The roots of the generator polynomial are in the "big field" $GF(2^m)$. On one hand, for encoding purposes it is sufficient to work only with the small field. On the other hand, decoding requires operations in the big field.

3.3.3 Decoding

The algebraic decoding of BCH codes can be carried out in four steps:

1. Compute the syndrome S
2. Determine an *error locator polynomial* (ELP), whose roots provide an indication of where the errors are. There are several different ways of finding the locator polynomial. These methods include Peterson's algorithm, and the

Fig. 9 Block diagram of the
standard BCH decoding
process

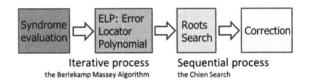

Berlekamp- Massey algorithm. In addition, there are techniques based upon
Galois-field Fourier transforms.
3. Find the roots of the error locator polynomial. This is usually done using the
   Chien search, which is an exhaustive search over all the elements in the field.
4. Correct the identified errors

These steps are outlined in Fig. 9.

Among the many decoding algorithms for BCH codes, Berlekamp's iterative
algorithm, and Chien search algorithm are the most efficient ones [16, 23]. However
these algorithms requires iterative operations, an then they are not suitable for being
embedded in memory interfaces with tight time constraints.

Suppose a codeword $\mathbf{v}$ is transmitted and $\mathbf{r}$ is the corresponding received n-tuple.
The syndrome of $\mathbf{r}$ is given by:

$$S = (S_1, S_3, \ldots, S_{2t-1}) = \mathbf{r} \cdot \mathbf{M}^T,$$

which actually consists of t syndrome components; each component S_i being an m-
tuple (an element in GF(2^m)). In Sect. 3.3.2 we have seen that, for binary codes,
$\mathbf{r}(\alpha^{2i}) = \mathbf{r}^2(\alpha^i)$, and then $S_{2i} = S_i^2$. This means that from the t syndromes
$(S_1, S_3, \ldots, S_{2t-1})$ it is possible to reconstruct the $2t$ syndromes
$(S_1, S_2, \ldots, S_{2t-1}, S_{2t})$.

Suppose now that the error pattern $\mathbf{e}$ has v errors at location $j_1, \ldots, j_v$, where
$0 \leq j_1 < j_2 < \cdots < j_v < n$. The syndrome S depends on the error pattern $\mathbf{e}$ only (not
on the received n-tuple $\mathbf{r}$) [16]. In particular the following relationship between
syndromes and error pattern holds true [16]:

$$S_i = \mathbf{e}(\alpha^i). \tag{6}$$

From the hypothesis on $\mathbf{e}$ and from (6) we obtain the following set of equations:

$$S_1 = \alpha^{j_1} + \alpha^{j_2} + \cdots + \alpha^{j_v}$$
$$S_2 = (\alpha^{j_1})^2 + (\alpha^{j_2})^2 + \cdots + (\alpha^{j_v})^2$$
$$\vdots$$
$$S_{2t-1} = (\alpha^{j_1})^{2t-1} + (\alpha^{j_2})^{2t-1} + \cdots + (\alpha^{j_v})^{2t-1}$$
$$S_{2t} = (\alpha^{j_1})^{2t} + (\alpha^{j_2})^{2t} + \cdots + (\alpha^{j_v})^{2t}$$

where $\alpha^{j_1}, \alpha^{j_2}, \ldots, \alpha^{j_v}$ are unknown. These equations are said to be *power-sum symmetric functions*. Any method for solving these equations is a decoding algorithm for the BCH codes.

Rather than attempting to solve these nonlinear equations directly, a new polynomial is introduced, the *error locator polynomial* (ELP), which casts the problem in a different—more tractable—setting. The error locator polynomial is defined as

$$\Lambda(x) = \prod_{l=1}^{v} \left(1 - \alpha^{j_l} x\right) = \lambda_0 + \lambda_1 x + \cdots + \lambda_{v-1} x^{v-1} + \lambda_v x^v$$

where $\lambda_0 = 1$. It's worth noting that, by this definition, $\Lambda(x) = 0$ if $x = a^{-j_l}$; that is, the roots of the error locator polynomial are at the reciprocals (in the field arithmetic) of the error locators.

The actual determination of the coefficients of the error locator polynomial is at the heart of any BCH decoding algorithm. A key point is that, while the power-sum symmetric functions provide a nonlinear relationship between the syndromes and the error locators, it can be shown [23] that there is a linear relationship between the syndromes and the coefficients of the error locator polynomial. This relation is called *key equation*:

$$S_i = -\sum_{l=1}^{v} \lambda_l S_{i-l}, \tag{7}$$

where $i = v+1, v+2, \ldots, 2v$. In matrix form, (7) can be rewritten as

$$
\begin{bmatrix}
S_1 & S_2 & \cdots & S_v \\
S_2 & S_3 & \cdots & S_{v+1} \\
\vdots & \vdots & \ddots & \vdots \\
S_{v-1} & S_v & \cdots & S_{2v-2} \\
S_v & S_{v+1} & \cdots & S_{2v-1}
\end{bmatrix}
\cdot
\begin{bmatrix}
\lambda_v \\
\lambda_{v-1} \\
\vdots \\
\lambda_2 \\
\lambda_1
\end{bmatrix}
=
\begin{bmatrix}
-S_{v+1} \\
-S_{v+2} \\
\vdots \\
-S_{2v-1} \\
-S_{2v}
\end{bmatrix}
\tag{8}
$$

As example, let $v = 1$—i.e., there is a single error – then (7) becomes

$$S_2 = \lambda_1 S_1.$$

In binary fields $S_2 = S_1^2$. Consequently we get that $\lambda_1 = S_1$, and the locator polynomially is $\Lambda(x) = 1 + S_1 x$. If the error is in position i, the condition $\Lambda(\alpha^{-i}) = 1 + S_1 \alpha^{-1} = 0$ implies $\alpha^i = S_1$. This means single-error correcting BCHs boil down to Hamming codes.

In general, BCH decoding algorithms exploit the set of equations in (8) to determine the error locator polynomial coefficients. For example, the Berlekamp-Massey algorithm [5, 11], iteratively forms the solution of (8), but any standard technique to invert a matrix can be applied to the task.

Once we have the error locator polynomial, the next step consists in finding its roots. In particular we need to find the error-location numbers that are the reciprocal of the roots of $\Lambda(x)$. The roots of $\Lambda(x)$ can be found by an exhaustive search based on substituting $1, \alpha, \alpha^2, \ldots, \alpha^{n-1}$ into $\Lambda(x)$. The Chien search algorithm [6] is a method of exhaustively testing the polynomial to find these roots, which when inverted, identify the error locations. Then, adding a one to their identified locations easily repairs the errors.

The received vector $\mathbf{r} = (r_0, \ldots, r_{n-1})$ is decoded bit by bit. The high order bits are decoded first. To decode r_{n-1}, the decoder test if α^{n-1} is an error-location number. This is equivalent to checking if its inverse α is a root of $\Lambda(x)$, i.e. checking if (Table 8)

$$1 + \lambda_1 \alpha + \lambda_2 \alpha^2 + \cdots + \lambda_\nu \alpha^\nu = 0.$$

Then the Chien search proceeds by evaluating $\Lambda(\alpha^2), \Lambda(\alpha^3), \ldots$, but instead of evaluating each term from scratch, the result of the previous step is utilized. To understand the principles of Chien search, let's look at the evaluation of $\Lambda(\alpha^l)$ and $\Lambda(\alpha^{l+1})$:

$$\Lambda(\alpha^l) = 1 + \lambda_1 \alpha^l + \lambda_2 \alpha^{2l} + \cdots + \lambda_\nu \alpha^{l\nu}$$
$$\Lambda(\alpha^{l+1}) = 1 + \lambda_1 \alpha^{l+1} + \lambda_2 \alpha^{2l+2} + \cdots + \lambda_\nu \alpha^{l\nu+\nu}.$$

We can see that the i-th term of $\Lambda(\alpha^{l+1})$ can be obtained by the i-th term of $\Lambda(\alpha^l)$ by multiplying that term by α^i.

If $\Lambda(\alpha^l) = 0$, then α^l is a root of Λ, and r_{n-l} is an erroneous bit. The correction step is achieved adding 1 to each bit in error.

3.3.4   Implementation of BM Decoding

A description of an efficient implementation of the standard BCH decoding flow was outlined by Strukov [28]. Here we report the key result, summarized in Fig. 8, where Step 1 is the syndrome evaluation, the Step 2 consists in finding the error-location polynomial using the Berlekamp-Massey iterative algorithm, and Step 3 consists in finding error-location numbers with subsequent error-correction. In particular, the Figure shows that for large values of n the area of the decoder scales approximately as nmt^2, and it is mostly dominated by the circuitry in Step 3. On the other hand, most of the delay, which is roughly proportional to mt, comes from Step 2.

The hypotheses behind this table are that m is small, and that the clock of the embedded microprocessor is not a bottleneck.

As example, let $m = 9$ and $t = 2$. By applying the equations in Fig. 8 and using the values in Table 1, we get that the decoder area is 31.24 k XOR, and its latency is 67.2 XOR levels.

Table 8 Complexity of BCH decoding solution based on Berlekamp-Massey algorithm. Taken from [28]

Operation	Area			Critical Path Delay		
	OR	AND	XOR	OR	AND	XOR
Step 1	0	0	tnm	0	0	$\lceil\log_2 n\rceil$
Step 2	$m^2 + nml/4$	$3m^2t + 2ml^2 + m^2n/4 - m^2$	$3m^2t + 2tm - m^2$	$t(m-1)$	$2t + t\lceil\log_2 m\rceil + 2t\lceil\log_2 t\rceil$	
Step 3	0	mn	$m^2tn/2 + n$	0	$\lceil\log_2 m\rceil$	$1 + \lceil\log_2 m\rceil + \lceil\log_2 t\rceil$
Total	$ml^2 + nm/4$	$3m^2t + 2ml^2 + m^2n/4 + mn - m^2$	$tnm + 3m^2t + 2tm + m^2tn/2 + n - m^2$	$t(m-1)$	$2t + t\lceil\log_2(m-1)\rceil + \lceil\log_2 m\rceil$	$1 + \lceil\log_2 n\rceil + (4t+1)\lceil\log_2 m\rceil + (2t+1)\lceil\log_2 t\rceil$

4 Ultra-Fast Double Error Correcting BCH Codes

This section is based on the ECC solutions developed by the authors for emerging memory applications [1, 2, 8].

To minimize the decoding latency, the decoder has to ben carefully designed. The iterative BM algorithm, usually used to manage NAND memories requiring ECC of high error correction capability, is not affordable with tight latency constraints. Instead, it can be replaced by the parallel evaluation of the ELP symbolic expressions, selected through a decision tree [14]. To make this possible, it is important to deal with a limited number of ELP expressions. Any time-consuming computation must be avoided whenever possible. When necessary, the evaluation of complex terms must be isolated and carried out as much as possible in parallel with the other terms, to avoid bottlenecks in the decoder.

Given the above precautions, the decoding algorithm is composed of the four classical stages shown in Fig. 9, namely Syndrome Evaluation, ELP computation, Exhaustive Search of the roots of the ELP, and final correction. The latency of these four steps is carefully optimized, both independently and jointly, as outlined in the next subsections.

The traditional decoding process of a BCH code starts from syndromes, computes the Error Locator Polynomial (ELP) by the iterative Berlekamp-Massey algorithm and searches the ELP roots, i.e. the error positions, using a Chien machine [16] (Fig. 9). To reduce the correction latency to a sfew nanoseconds the decoding algorithm has been rewritten from scratch removing iterative and sequential processes. For this reason the ELP coefficients are expressed in terms of syndrome avoiding the division in the Galois Field, a computationally heavy operation, and searching the roots by trying all the possibilities simultaneously.

The minimum Hamming distance of a binary code $\mathscr{C}$ able to correct $t = 2$ errors is $d = 5$. The generator polynomial of a BCH code designed in $GF(2^m)$ with distance 5 has 4 consecutive roots in the exponential representation of α^i, plus the conjugate roots [16]

$$g(x) = \prod_{i=0}^{m-1} \left(x - \alpha^{2^i}\right)\left(x - \alpha^{3 \cdot 2^i}\right). \tag{9}$$

Some cyclotomic cosets may include less than m items. In that case the degree of $g(x)$ is lower than $2\,m$ as assumed in (9).

Usually in a memory device the information data size k is a power of 2 (e.g., 256 bits). Let's suppose that $k = 2^{m-1}$. Then to encode a message $\mathbf{u}$ of k bits we need a BCH code designed in a larger field, for example in $GF(2^m)$. The final code $\mathscr{C}$ can be obtained by shortening a primitive code of length $N = 2^m - 1$, with $N - K \leq mt$ parity bits, K information bits and generator polynomial $g(x)$ with roots in $GF(2^m)$.

As an example let $k = 16$, then $m = 5$ and the generator polynomial with roots in $GF(2^5)$ is

$$\mathbf{g}(x) = 1 + x^3 + x^5 + x^6 + x^8 + x^9 + x^{10}.$$

Consequently the starting primitive BCH2 is a $(N, K) = (31, 21)$ code with ten parity-check bits. Since $K = 21$, the shortened code with $k = 16$ is obtained by removing five information bits. Thus the code $\mathscr{C}$ is a (26,16) BCH2.

Let's apply the shortening by keeping the first k information bits. Then the 10×26 parity-check matrix of this code is

$$
\mathbf{M} =
\left[
\begin{array}{cccccccccccccccccccccccccc}
1 & 0 & 0 & 0 & 0 & 1 & 0 & 0 & 1 & 0 & 1 & 1 & 1 & 1 & 0 & 0 & 0 & 1 & 1 & 0 & 1 & 1 & 1 & 0 & 1 & 0 \\
0 & 1 & 0 & 0 & 0 & 0 & 1 & 0 & 0 & 1 & 1 & 1 & 1 & 1 & 1 & 0 & 0 & 0 & 1 & 1 & 0 & 1 & 1 & 1 & 0 & 1 \\
0 & 0 & 1 & 0 & 0 & 1 & 0 & 1 & 1 & 0 & 1 & 0 & 0 & 0 & 1 & 1 & 0 & 1 & 1 & 1 & 0 & 1 & 0 & 1 & 0 & 0 \\
0 & 0 & 0 & 1 & 0 & 0 & 1 & 0 & 1 & 1 & 1 & 1 & 0 & 0 & 0 & 1 & 1 & 0 & 1 & 1 & 1 & 0 & 1 & 0 & 1 & 0 \\
0 & 0 & 0 & 0 & 1 & 0 & 0 & 1 & 0 & 1 & 1 & 1 & 1 & 0 & 0 & 0 & 1 & 1 & 0 & 1 & 1 & 1 & 0 & 1 & 0 & 1 \\
\hline
1 & 0 & 0 & 0 & 0 & 1 & 1 & 0 & 0 & 1 & 1 & 1 & 0 & 1 & 1 & 1 & 0 & 0 & 0 & 1 & 0 & 1 & 0 & 1 & 1 & 0 \\
0 & 0 & 1 & 1 & 1 & 1 & 1 & 0 & 1 & 1 & 0 & 1 & 0 & 1 & 1 & 0 & 1 & 0 & 0 & 0 & 0 & 1 & 1 & 0 & 0 & 1 \\
0 & 0 & 0 & 0 & 1 & 1 & 0 & 0 & 1 & 0 & 1 & 0 & 1 & 1 & 1 & 0 & 0 & 0 & 1 & 0 & 1 & 0 & 1 & 1 & 0 & 1 \\
0 & 1 & 1 & 1 & 1 & 1 & 0 & 1 & 1 & 1 & 1 & 0 & 1 & 1 & 0 & 1 & 0 & 0 & 0 & 0 & 1 & 1 & 0 & 0 & 1 & 0 \\
0 & 0 & 0 & 1 & 0 & 1 & 0 & 1 & 1 & 0 & 1 & 1 & 0 & 0 & 1 & 0 & 0 & 1 & 1 & 1 & 1 & 1 & 0 & 1 & 1 & 1 \\
\end{array}
\right],
$$

and the associated 16×26 generator matrix $\mathbf{G}$ is

$$
\mathbf{G} =
\left[
\begin{array}{cccccccccc}
1 & 0 & 0 & 1 & 0 & 1 & 1 & 0 & 1 & 1 \\
1 & 1 & 0 & 1 & 1 & 1 & 0 & 1 & 1 & 0 \\
0 & 1 & 1 & 0 & 1 & 1 & 1 & 0 & 1 & 1 \\
1 & 0 & 1 & 0 & 0 & 0 & 0 & 1 & 1 & 0 \\
0 & 1 & 0 & 1 & 0 & 0 & 0 & 0 & 1 & 1 \\
1 & 0 & 1 & 1 & 1 & 1 & 1 & 0 & 1 & 0 \\
0 & 1 & 0 & 1 & 1 & 1 & 1 & 1 & 0 & 1 \\
1 & 0 & 1 & 1 & 1 & 0 & 0 & 1 & 0 & 1 \\
1 & 1 & 0 & 0 & 1 & 0 & 1 & 0 & 0 & 1 \\
1 & 1 & 1 & 1 & 0 & 0 & 1 & 1 & 1 & 1 \\
1 & 1 & 1 & 0 & 1 & 1 & 1 & 1 & 0 & 0 \\
0 & 1 & 1 & 1 & 0 & 1 & 1 & 1 & 1 & 0 \\
0 & 0 & 1 & 1 & 1 & 0 & 1 & 1 & 1 & 1 \\
1 & 0 & 0 & 0 & 1 & 0 & 1 & 1 & 0 & 0 \\
0 & 1 & 0 & 0 & 0 & 1 & 0 & 1 & 1 & 0 \\
0 & 0 & 1 & 0 & 0 & 0 & 1 & 0 & 1 & 1 \\
\end{array}
\;\middle|\; \mathbf{I}_{16}
\right].
$$

4.1   Elementary Operations in $\mathrm{GF}(2^m)$

In this section we describe the full-parallel implementation of elementary operations in $\mathrm{GF}(2^m)$.

4.1.1 Multiplication by a Constant

Let be a be a variable in $GF(2^m)$, and α^j a constant. The product $b = a\alpha^i$ can be written as

$$b = a\alpha^i = \sum_{i=0}^{m-1} a_i \alpha^i \alpha^j = \sum_{i=0}^{m-1} a_i \alpha^{i+j}.$$

and, in matrix form, as

$$b = (a_0, \ldots, a_{m-1}) \cdot \begin{bmatrix} \alpha^j \\ \alpha^{j+1} \\ \vdots \\ \alpha^{j+m-2} \\ \alpha^{j+m-1} \end{bmatrix}$$

Thus it is enough to know the binary representation of $\alpha^j, \alpha^{j+1}, \ldots, \alpha^{j+m-1}$ to identify the linear combination of the bits of a in order to obtain the m bits of b.

As example let's consider $GF(2^5)$, and take α^6 as constant. The binary representation of $GF(2^5)$ elements are in Table 7. In this example

$$b = a\alpha^6$$

$$= (a_0, a_1, a_2, a_3, a_4) \cdot \begin{bmatrix} \alpha^6 \\ \alpha^7 \\ \alpha^8 \\ \alpha^9 \\ \alpha^{10} \end{bmatrix}$$

$$= (a_0, a_1, a_2, a_3, a_4) \cdot \begin{bmatrix} 0 & 1 & 0 & 1 & 0 \\ 0 & 0 & 1 & 0 & 1 \\ 1 & 0 & 1 & 1 & 0 \\ 0 & 1 & 0 & 1 & 1 \\ 1 & 0 & 0 & 0 & 1 \end{bmatrix}$$

Now let $a = \alpha^7$, then $a\alpha^6 = \alpha^7\alpha^6 = \alpha^{13}$. By multiplying the binary representation of $\alpha^7 = (0, 0, 1, 0, 1)$ by the above matrix we get $(0,0,1,1,1)$ which is the binary representation of α^{13}.

We have seen that the multiplication by a constant is computed by using an $m \times m$ matrix. On average each column is half filled with ones. Hence the implementation of this operation requires m Xor trees, each using $m/2 - 1$Xor s and with average depth $\log_2(m/2)$.

4.1.2 Multiplication of Two Variables

Various structures have been proposed to compute $c = ab$, with $a, b \in \mathrm{GF}(2^m)$. Here we adopt the *Mastrovito multiplier* [21], that exploits a possible different latency between the two factors a and b. We can write $c = ab$ as

$$c = ab = a \sum_{j=0}^{m-1} b_j \alpha^j = \sum_{j=0}^{m-1} b_j (a\alpha^j).$$

The products

$$a\alpha^j = \sum_{i=0}^{m-1} a_i^{(j)} \alpha^i,$$

where $j = 0, \ldots, m-1$, can be prepared in advance. The m products and $m-1$ additions in

$$c_i = \sum_{j=0}^{m-1} b_j a_i^{(j)}.$$

start when also b is available. The total latency depends on the binary representation of the powers α^j to α^{m-1+j}, with $j = 1, \ldots, m-1$ thus on the field.

For instance in $\mathrm{GF}(2^9)$ the maximum latency for the terms $a_i^{(j)}$ is $2T_X$. However some terms require only T_X and can be multiplied in advance by the corresponding b_j (if available). With this precaution, the computation of each bit c_i can be completed within $3T_X$, despite nine addends, and the total latency of the operation is $T_A + 5T_X$ (with a maximum allowed extra delay T_X for b).

Even an additional sum, i.e., $ab + d$ or $ab + d^2$ or even $ab + d^4$, can be completed in parallel during the computation of ab.

An equivalent implementation can be done also for $c = a^2 b$. Powers a^{2^k} (see the next subsection) are linear combinations of a whose evaluation can be embedded in the terms $a^{2^k} \alpha^j$ with no additional delay.

4.1.3 Powers

The computation of powers as $c = a^{2^k}$ is simple, as they require only linear combinations of the bits a_i for each bit c_j. Let's start by considering the square operation $c = a^2$. We have

$$c = a^2 = \left(\sum_{i=0}^{m-1} a_i \alpha^i \right)^2 = \sum_{i=0}^{m-1} a_{2i} \alpha^{2i} = \sum_{i=0}^{m-1} a_{2i} \alpha^{2i}$$

and in matrix form:

$$a^2 = (a_0, \ldots, a_{m-1}) \cdot \begin{bmatrix} 1 \\ \alpha^2 \\ \vdots \\ \alpha^{2(m-2)} \\ \alpha^{2(m-1)} \end{bmatrix}$$

As example let $a = \alpha^{11} \in GF(2^5)$ the above equation in binary representation becomes:

$$(\alpha^{11})^2 = (1,1,1,0,0) \cdot \begin{bmatrix} 1 & 0 & 0 & 0 & 0 \\ 0 & 0 & 1 & 0 & 0 \\ 0 & 0 & 0 & 0 & 1 \\ 0 & 1 & 0 & 1 & 0 \\ 1 & 0 & 1 & 1 & 0 \end{bmatrix} = (1,0,1,0,1) = \alpha^{22}$$

The latency depends on the actual field. For example, in $GF(2^9)$ α^2 requires one single level of XOR, whereas α^4 requires $2T_X$.

The computation of powers $c = a^n$ with $n \neq 2^k$ is complicated by the fact that non linear terms are required. To minimize the latency we can separate a linear part from a non-linear one as done, e.g., for a^3 in [1]:

$$
\begin{aligned}
c &= a^3 \\
&= \left(\sum_{i=0}^{m-1} a_i \alpha^i \right)^3 \\
&= \underbrace{\sum_{i=0}^{m-1} a_i \alpha^{3i}}_{\text{linear}} + \underbrace{\sum_{i=0}^{m-2} \sum_{j=i+1}^{m-1} a_i a_j (\alpha^{2i+j} + \alpha^{i+2j})}_{\text{non-linear}},
\end{aligned}
\tag{10}
$$

where on the left it is possible to identify the sum of cubes (linear operations), and on the right products of a bits. The binary representation of the $\alpha^{2i+j} + \alpha^{i+2j}$ can be computed from Tables 6 and 7.

The products $a_i a_j$ requires $m(m-1)/2$ AND s, and a single AND level. The vector of all these products can then be multiplied by the following $m(m-1)/2 \times m$ matrix

$$(a_0 a_1, a_0 a_2, \ldots, a_{m_3} a_{m-2}, a_{m-2} a_{m-1}) \cdot \begin{bmatrix} \alpha + \alpha^2 \\ \alpha^2 + \alpha^4 \\ \vdots \\ \alpha^{3m-8} + \alpha^{3m-7} \\ \alpha^{3m-5} + \alpha^{3m-4} \end{bmatrix}$$

Table 9 Summary of the computation latency (and allowed delays of the various terms) for the operations implemented in GF(2^9)

Operation	Latency	$b$delay	$c$delay
a^2	T_X		
a^4	$2T_X$		
ab	$T_A + 5T_X$	T_X	
$ab + c^2$	$T_A + 5T_X$	T_X	T_X
$a^2b + c$	$T_A + 5T_X$	T_X	$3T_X$
$a^3 + b$	$T_A + 4T_X$	$3T_X$	
a^6	$T_A + 4T_X$		

On the average the columns of the above matrix are half filled with ones. Then the computation of the above product requires m XOR trees (one for each column) with $m(m - 1)/4$ inputs. Thus the total number of XOR is $m(m(m - 1)/4 - 1)$, and the maximum depth of the tree is $\log_2(m(m - 1)/2)$.

In GF(2^9) by carefully optimizing the sequence and order of sums and products, the computation of a^3 can be completed within $T_A + 4T_X$. A similar implementation, with the same latency, can be done also for a^6.

As to the computation of $c = ab^3$, the minimum latency is achieved computing b^3 as described above, and the terms $a\alpha^j(j = 0...m - 1)$ in the meantime. The products require a single level of m^2 AND, and the final sum of products can be optimized. In a similar manner, the evaluation of terms like $ab^3 + d + e + ...$ can be realized without additional latency (Table 9).

4.2 Syndrome Evaluation

Starting from the received sequence $\mathbf{r} = (r_0, ...r_{n-1})$, with $r_i \in GF(2)$, the syndromes S_1 and $S_3 \in GF(2^m)$ can be computed by

$$S_1 = \mathbf{r}(\alpha) = \sum_{i=0}^{n} r_i\alpha^i$$

$$S_3 = \mathbf{r}(\alpha^3) = \sum_{i=0}^{n} r_i\alpha^{3i}$$

These operations can be rewritten in matrix form as

$$\begin{bmatrix} S_1 \\ S_3 \end{bmatrix} = \mathbf{r}\mathbf{M}^T = \mathbf{r}\begin{bmatrix} \mathbf{M}_1 \\ \mathbf{M}_3 \end{bmatrix}^T \tag{11}$$

where $\mathbf{M}$ is the parity-check matrix defined in (5).

4.3   Error-Locator-Polynomial Analysis

When two errors occur ($v = 2$), say in positions i_1, i_2, the ELP with roots in α^{-i_j} reads

$$\Lambda(x) = \left(1 - x\alpha^{i_1}\right)\left(1 - x\alpha^{i_2}\right) = 1 + \lambda_1 x + \lambda_2 x^2.$$

The coefficients of $\Lambda(x)$ can be evaluated by running the BM algorithm in symbolic form as in [14]:

$$\lambda_1 = S_1$$

$$\lambda_2 = \frac{S_1^3 + S_3}{S_1}$$

and the resulting error locator polynomial is then

$$\Lambda(x) = 1 + S_1 x + \frac{S_1^3 + S_3}{S_1} x^2$$

The actual values of the coefficients can be computed once S_1 and S_3 are available. The computation of λ_2 in this form is too time-consuming, since among the elementary operations the division is the most demanding one and must be avoided. To this aim $\Lambda(x)$ can be multiplied by S_1, obtaining an equivalent (i.e. with the same roots) polynomial:

$$\Lambda(x) = S_1 + S_1^2 x + (S_1^3 + S_3)x^2 \tag{12}$$

For each $i = 0, \ldots, n - 1$ we have to check whether $\Lambda(\alpha^i) = 0$. If yes, the i-th bit is in error and the correction is applied by inverting it.

Let i be the error position. Then checking $\Lambda(\alpha^{-i}) = 0$ means evaluating one of the following conditions:

$$S_1 + S_1^2 \alpha^{-i} + (S_1^3 + S_3)\alpha^{-2i} = 0$$
$$S_1 \alpha^i + S_1^2 + (S_1^3 + S_3)\alpha^{-i} = 0$$
$$S_1 \alpha^{2i} + S_1^2 \alpha^i + S_1^3 + S_3 = 0$$

These conditions are equivalent because $\alpha^i \neq 0$. Among them, the more useful is the last one, because the most complex computation is that of $S_1^3 + S_3$, and is the last term that becomes available.

The presence of two errors is indicated by the conditions $S_1 \neq 0$ and $S_1^3 + S_3 \neq 0$. If instead $S_1 \neq 0$ and $S_1^3 + S_3 = 0$, the error locator polynomial is

$$\Lambda(x) = 1 + S_1 x.$$

Again by multiplying by S_1 we get a polynomial with the same roots:

$$\Lambda(x) = S_1 + S_1^2 x.$$

This means that the error locator polynomial for a single error is included as particular case of the double error error locator polynomial (12).

At last, if $S_1 = 0$ there isn't any error, or there are at least three errors. Since the error locator polynomial in (12) has been multiplied by S_1, its evaluation would be 0 for any input. Thus the correction has to be explicitly disabled. Finally we remark that if $S_1 = 0$ and $S_3 \neq 0$ the error locator polynomial has degree two, but it has no roots. Then the only case in which the correction must be disabled is when $S_1 = 0$ and $S_3 = 0$.

4.4 Decoder Architecture

In this section we analyze the overall architecture of the proposed decoder shown in Fig. 10.

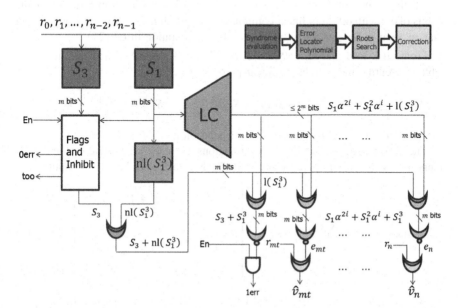

Fig. 10 Logical scheme of the decoder of the 2-error-correcting BCH code

4.4.1 Syndrome Evaluation

The first step is realized by the blocks S_1 and S_3 in Fig. 10. These blocks implement the linear combinations (11) of the received vector **y** producing the tow m-bit syndromes S_1 and S_3 in GF(2^m).

In a completely bit-parallel implementation each bit of a syndrome S_i can be obtained by a separate XOR-tree with inputs taken from the received code vector [15].

On the average, the binary columns of $\mathbf{M}_i$ ($n \times m$) are half filled with ones while in the worst case there is a column with almost all ones. Hence the syndrome S_i calculation circuit will be comprised of a total of m XOR-trees requiring on average $\left(\frac{1}{2}n - 1\right)$ XOR each.

The average depth is $\log_2(n/2)$ and the worst depth is $\log_2(n)$.

4.4.2 Error Locator Polynomial

We have previously seen how to write the ELP without divisions in the GF in such a way to reduce the number of stages for its evaluation when checking a potential error position. The ELP evaluation is structured into a sum of two terms: one linear and one nonlinear in the components of the syndrome vector. The linear component —the block **LC** in Fig. 10—is bit-position dependent and all possible values for all the positions are evaluated in parallel. The nonlinear component, generated by the third power of the syndrome vector S_1—the block $\mathbf{S_1^3}$ NL in Fig. 10—is the most time consuming so that a preferential path is set for it to prevent additional delays.

All the m bits of $S_1\alpha^{2i} + S_1^2\alpha^i$ can be computed as linear combinations of the m bits of S_1; with different linear combinations for each i. This follows from the fact that, as seen in Sect. 4.1.3, squaring is a linear operation in GF(2^m).

From (10), even for $S_1^3 = (S_{1,0} + S_{1,1}\alpha + S_{1,2}\alpha^2 + \cdots + S_{1,m-1}\alpha^{m-1})^3$, it is possible to identify and extract the linear part

$$\lin(S_1^3) \triangleq S_{1,0} + S_{1,1}\alpha^3 + S_{1,2}\alpha^6 + \cdots + S_{1,m-1}\alpha^{3(m-1)}.$$

These linear combinations does not depend on i.

The coefficients of the linear combinations for each of k values of i can be identified with simple operations in GF(2^m). For each i we have to compute

$$S_1\alpha^{2i} + S_1^2\alpha^i + \sum_{j=0}^{m-1} S_{1,j}\alpha^{3j} = S_1 \cdot \begin{bmatrix} \alpha^{2i} + \alpha^i + 1 \\ \alpha^{2i+1} + \alpha^{i+2} + \alpha^3 \\ \vdots \\ \alpha^{2i+m-2} + \alpha^{i+2(m-2)} + \alpha^{3(m-2)} \\ \alpha^{2i+m-1} + \alpha^{i+2(m-1)} + \alpha^{3(m-1)} \end{bmatrix}. \tag{13}$$

As example let take GF$\left(2^5\right)$, and let $i = 6$, and $S_1 = \alpha^7 = (0,0,1,0,1)$.

$$S_1 \cdot \begin{bmatrix} \alpha^5 \\ \alpha^{25} \\ \alpha^{19} \\ \alpha^{27} \\ \alpha^3 \end{bmatrix} = (0,0,1,0,1) \cdot \begin{bmatrix} 1 & 0 & 1 & 0 & 0 \\ 1 & 0 & 0 & 1 & 1 \\ 0 & 1 & 1 & 0 & 0 \\ 1 & 1 & 0 & 1 & 0 \\ 0 & 0 & 0 & 1 & 0 \end{bmatrix} = (0,1,1,1,0) = \alpha^{12}$$

For each i the needed linear combinations are obtained by looking at the columns of the binary matrix in (13). In the example they are

$$S_{1,0} + S_{1,0} + S_{1,3}$$
$$S_{1,2} + S_{1,3}$$
$$S_{1,0} + S_{1,2}$$
$$S_{1,1} + S_{1,3} + S_{1,4}$$
$$S_{1,2}$$

For the sake of brevity, we can represent them with their associated integer:

$$11, 12, 5, 26, 2$$

The following matrix represent the linear combinations needed by all the bits of the code (26,16):

$$\begin{bmatrix} 17 & 13 & 11 & 7 & 0 & 27 & 11 & 10 & 29 & 10 & 16 & 23 & 1 & 12 & 28 & 16 & 26 & 28 & 29 & 7 & 13 & 6 & 26 & 17 & 0 & 22 \\ 22 & 29 & 28 & 13 & 20 & 22 & 12 & 20 & 7 & 4 & 30 & 23 & 12 & 5 & 15 & 14 & 14 & 31 & 23 & 29 & 13 & 5 & 30 & 6 & 4 & 15 \\ 6 & 21 & 3 & 0 & 29 & 21 & 5 & 14 & 5 & 8 & 11 & 16 & 22 & 30 & 8 & 13 & 30 & 14 & 3 & 6 & 19 & 13 & 24 & 0 & 27 & 27 \\ 14 & 30 & 6 & 26 & 27 & 22 & 26 & 3 & 18 & 31 & 11 & 22 & 2 & 7 & 23 & 23 & 15 & 11 & 14 & 6 & 2 & 31 & 19 & 18 & 7 & 15 \\ 28 & 20 & 19 & 30 & 7 & 7 & 2 & 28 & 17 & 13 & 19 & 27 & 25 & 10 & 30 & 2 & 25 & 15 & 0 & 15 & 5 & 17 & 8 & 13 & 22 & 5 \end{bmatrix}$$

where the combinations for th i-th bit, with $i = 0, \ldots, n-1$, are found in the i-th column.

We have k different positions and m (possibly) different linear combinations for each of these. Thus in total mk combinations would be needed. Even in the case that $k = 2^{m-1}$, we have that $mk \geq 2^m$ (if $m \geq 2$). Hence it is worthwhile to evaluate all the 2^m distinct linear combinations of m bits, and select the appropriate one for each term and position.

To evaluate all the possible linear combinations of m bits we need $2^m - 1$ XOR trees. The total number of XOR is:

$$\sum_{i=1}^{m} \binom{m}{i}(i-1) = \sum_{i=1}^{m} \binom{m}{i} i - \sum_{i=1}^{m} \binom{m}{i} = m2^{m-1} - (2^m - 1),$$

and the maximum depth of the trees is $\lceil \log_2 m \rceil$ XOR levels.

The non-linear part of S_1^3 is

$$\mathrm{nl}(S_1^3) \triangleq \sum_{i=0}^{m-2} \sum_{j=i+1}^{m-1} S_{1,i} S_{1,j} (\alpha^{2i+j} + \alpha^{i+2j}),$$

and it is exactly the non-linear part of the cube operation given in (10), whose implementation is described in Sect. 4.1.3.

The non-linear part $\mathrm{nl}(S_1^3)$ is then added to the syndrome S_3. This requires $m\mathrm{XOR}$, and a single XOR level.

4.4.3 Root Search

Let A_i be the linear combination for the i-th bit, and B be the sum of the S_3 with the non-linear part of S_1^3 and B be the sum of the S_3 with the non-linear part of S_1^3:

$$A_i \triangleq S_1 \alpha^{2i} + S_1^2 \alpha^i + 1(S_1^3)$$
$$B \triangleq \mathrm{nl}(S_1^3) + S_3$$

For each bit i the root search computes $A_i + B$. If $A_i + B = 0$ the i-th bit is in error, and it must be inverted. This is obtained by NORing the bits $A_i + B$, and XORing the result with the bit r_i:

$$\hat{v}_i = r_i + \mathrm{NOR}_m(A_i + B)$$

where NOR_m gate is a NOR gate with m inputs. Such a gate can be implemented by using a tree of elementary NOR and NAND gates. See Appendix "Counting the Number of Elementary Gates in a n-Input Gate".

4.4.4 Estimates

To have a first-order approximation of the area occupancy as a function of m, it is possible to use the formulae in Table 10. These formulae tend to over-estimate the

Table 10 Theoretical estimates of area occupancy for the BCH2 decoder blocks implemented in GF(2^m))

Block	Area	Latency
Syndrome	$2m(n/2 - 1)\mathrm{XOR}$	$(m - 1)\mathrm{XOR}$
ELP	$(m^3/4 - m^2/4 + 2^{m-1}m - 2^m + 1)$ $\mathrm{XOR} + m(m - 1)/2\mathrm{AND}$	$(\lceil \log_2(m(m - 1)/2) \rceil + 1)\mathrm{XOR} + 1\mathrm{AND}$
Correction	$k(m + 1)\mathrm{XOR} + k\,\mathrm{NOR}_m$	$2\mathrm{XOR} + 1\mathrm{NOR}_m$
Total	Synd + Linear + Nonlinear + Corr	Synd + Max(Linear,Nonlinear) + Corr

actual area, because they consider just the values of n, k and m (and not completely the actual matrices content) and do not take into account any optimizations related to a specific Galois field. The cost in area of a NOR_m gate and the Iverson's notation $[\lceil \log_2(m) \rceil \text{ even}]$ are described in the Appendix.

As an example, let $m = 9$ and $t = 2$ (n is then 274). By applying the equations in Table 10 and using the values in Table 1, we get that the decoder area is 7.6kXOR, and its latency is 18.4XOR levels. By comparing this with the solution described in Sect. 3.3.4, the area is reduced by 77% and the latency by 73%. As we will see in the next section, the gain can be further increased by making some optimizations which are specific of the selected Galois field—here $GF(2^9)$.

5   Example of Implementation on PCM Devices

The BCH2 architecture described in the previous section has been implemented in a 45 nm 1G bit PCM device [31], which went into mass production in 2010 and sold millions of units. Such an implementation was able to correct up to two errors over 256 data bits with no more than 10 ns of overhead latency with an 180 nm CMOS logic gate length.

The shortened (274, 256) BCH code is obtained deleting (setting to logical 0) 237 selected data bits. In the following $i_n \in \{0, 1, \ldots, 510\}$, with $n \in \{0, 1, \ldots, 273\}$, indicates the degree of the nth element of the shortened codewords. The choice of 256 data bits out of 493 gives many degrees of freedom, which can be spent to achieve particular features for the final code. For instance the choice can be done in order to include the all-ones pattern in the shortened code thus allowing to the bit-wise NOT of a generic codeword to become a codeword itself. In this way the all-ones pattern (equivalent to the erased state of flash memories) is covered by the ECC, and moreover an energy reduction policy based on pattern inversion is implementable. The remaining degrees of freedom should be used to choose in the original parity check matrix H' of size 18×511, the 256 information positions in such a way that one parity bit is always zero in order to reduce also the number of parity check bits. Unfortunately the degrees of freedom are not enough to achieve this feature. In any case, it is possible to use the residual degrees of freedom to minimize the weight of the parity matrix overall (thus minimizing the decoding area) and the maximum weight of its columns (thus minimizing the decoding delay). The final code $\mathscr{C}$ has 18 parity bits in addition to 256 data bits, i.e. it is a shortened (274, 256) BCH code. Figure 11 describes the shortening stage, and Fig 12 shows the main functional blocks of the resulting code.

Because the device has the read-while-write feature, two ECC decoder machines are placed in order to have two independent data path simultaneously active, one for read and one for write (verify). The ECC encoder is placed once, in the write path only.

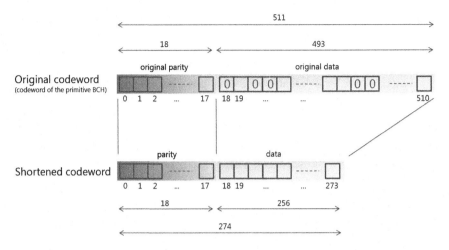

Fig. 11 Representation of the shortening

From the previous analysis it follows that the most critical path, in terms of latency, is the evaluation of the nonlinear part of S_1^3 that requires 1 AND + 4 XOR levels, after S_1 is available, i.e., 1 AND + 11 XOR after reading. We chose to sum S_3 at this stage instead of adding it to each variable term $S_1\alpha^{2i} + S_1^2\alpha^i$, as it would have saved a little time (1 XOR) with a high cost in terms of gates (2295 extra XOR). The final correction stage requires the addition of the constant to the variable term (1 XOR level), the check of the sum being null (2 OR$_3$ levels, where OR$_3$ is a 3-input-OR) and the correction (1 XOR level) for a total correction time of

$$T_{\text{correction}} = 1\,\text{AND} + 14\,\text{XOR} + 2\text{OR}_3 \tag{14}$$

that in terms of basic elementary gates (INV, NAND, NOR, XOR) becomes

$$T_{\text{correction}} = 1\,\text{INV} + 3\,\text{NAND} + 2\,\text{NOR} + 14\,\text{XOR} \tag{15}$$

By using the values in Tab. 1, the total latency is 16.2 XOR levels. This is a further reduction of around 12% with respect to the generic ultra-fast decoding solution described in Sect. 4, and a 76% reduction with respect to the solution described in Sect. 3.3.4.

As to the area occupancy, the total number of equivalent XOR gates is around 6 k.

5.1 Effective Consumption, Time and Area Overhead

Special attention has been paid to avoiding undesirable consumption. The syndrome feed on very complex combinatorial logic. Glitches in syndromes or delays in the propagation of its components may have a large, undesired consumption effect because of the avalanche amplification of switching. A latching stage on the syndrome may be useful to prevents undesirable power consumption.

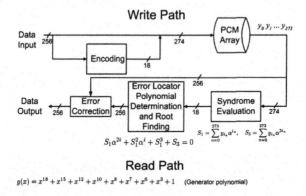

Fig. 12 Main functional blocks of the two-bit BCH solution

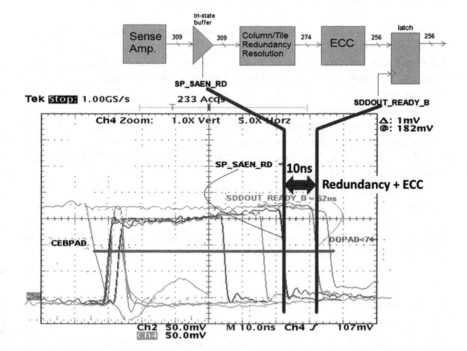

Fig. 13 Access time break-down highlighting the impact of ECC decoding

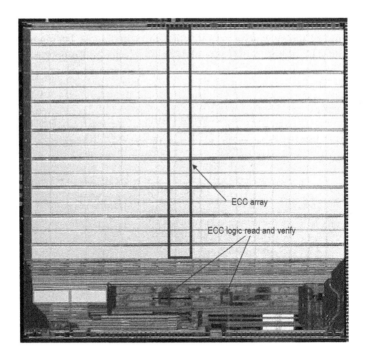

Fig. 14 Die micrograph

From the results of simulations the time overhead due to the critical path previously estimated, i.e. $T_{correction}$ in (14), is 10 ns in the worst case condition (1.6 V, 105 C and slow silicon). Figure 13 shows the signals involved in the entire reading sequence, and highlights the low impact of the ECC decoding integrated with the on board redundancy replacement.

Considering that the logic synthesis was performed using a library with a nominal density of 80 Kgate/mm^2, the total area cost is the 7% (18/256) of the array used to store the parity plus the encoding/decoding logic of 20 Kgate corresponding to 0.250 mm^2. Actually this logic were duplicated to enable the read-while-write feature of the device.

Figure 14 shows the die micrograph together with an indication of the regions reserved for parity bits and ECC logic.

By using high speed CMOS logic as in current DRAM technology nodes, the decoding time can be easily scaled down to 1 ns, thus making the 2-bit-corrector virtually transparent to the user.

6 Conclusions

ECC solutions are being used in practically all memory and storage devices. Since the latency of NAND flash memories is in the order of µs, the major focus of seeking ECC solutions for them has been optimization of the throughput. But when we consider and storage class memory devices, we are dealing with access time on the order of ns (for example a typical access time of current DRAM—devices is around 50 ns). Consequently, the latency of the ECC decoder also plays a critical role in such applications.

In this chapter we have considered ECC solutions suitable for the latter class of devices: SPC, Hamming and double-error-correcting codes. We have shown that by parallelizing the decoding algorithm, we are able to achieve low-latency decoding (even less than 1 ns). Moreover we have shown also that it is possible to achieve an high level of parallelism without an exponential increase of the ECC logic area.

Acknowledgements We gratefully acknowledge the entire EM Design Team, and the R&D EM Technology Development Team of Micron.

A special thanks goes to Sandro Bellini, Marco Ferrari, and Alessandro Tomasoni from Politecnico di Milano, with whom we have been working for years in exploring error correcting codes for memory applications. This works stems from the collaboration with them.

Appendix

Counting the Number of Elementary Gates in a n-Input Gate

An n-input gate can be implemented by using a binary-tree of 2-input elementary gates. For the sake of simplicity, let n be a power of 2 ($n = 2^m$), and as example, let's consider the $NAND_8$ gate. This 8-input gate can be implemented with a tree of depth 3 such that

- first stage: 4 $NAND_2$ gates (leaves of the tree)
- second stage: 2 NOR_2 gates
- third stage: 1 $NAND_2$ (the root of the tree)

The key point is that when $NAND_2$ and NOR_2 gates are used, the tree contains alternatively at each height one the 2 kinds of gate. NOR_m starts with NOR_2 as leaves, while $NAND_m$ starts with $NAND_2$ as leaves. Considering the $NAND_{16}$ gate, we have a tree of depth 4:

- first stage: 8 $NAND_2$ gates (leaves of the tree)
- second stage: 4 NOR_2 gates
- third stage: 2 $NAND_2$ gates
- fourth stage: 1 OR_2 (the root of the tree)

The cases $n = 8$ and $n = 16$ show that if m is odd, the root-gate is negative ($NAND_2$ or NOR_2), if m is even, the root-gate is direct (AND_2 or OR_2). In conclusion, the equivalent tree of the original negative $n = 2^m$-input gate ($NAND_n$ or NOR_n) is composed of elementary negative input gates ($NAND_2$ or NOR_2) only, except the case of m even in which the tree is followed by an inverter.

Now let n be any integer (not only a power of 2), and let $L = \lceil \log_2 n \rceil$. Let n_1 be the number of gates of the type used for the leaves of the tree, and n_{nl} the number of gates of the other type. Let n_{inv} the number of inverters (if needed). Thus $n_l + n_{nl} = n - 1$.

$$n_1 = \sum_{\substack{1 \le i < L \\ i \text{ odd}}} \left\lfloor \frac{n}{2^i} \right\rfloor$$

$$n_{nl} = n - 1 - n_1 \tag{16}$$

$$n_{inv} = [L \text{ even}]$$

where $[P(X)]$ is the Iverson's convention [9]. If $P(X)$ is true the term $[P(X)]$ is 1; otherwise it is 0.
The critical path is composed of

$$\frac{L + [L \text{ odd}]}{2} {}''\text{leave}'' \text{ gate} + \frac{L - [L \text{ odd}]}{2} {}''\text{non} - \text{leave}'' \text{ gate} + [L \text{ even}]\text{INV}$$

References

1. Amato, P., Bellini, S., Ferrari, M., Laurent, C., Sforzin, M., Tomasoni, A.: Ultra fast, two-bit ECC for emerging memories. In: Proc. of 6th IEEE International Memory Workshop (IMW), pp. 79–82. Taipei, Taiwan (2014)
2. Amato, P., Bellini, S., Ferrari, M., Laurent, C., Sforzin, M., Tomasoni, A.: Fast decoding ECC for future memories. IEEE Journal on Selected Areas in Communications **34**(9), 1–12 (2016). DOI 10.1109/JSAC.2016.2603698
3. Arikan, E.: Channel polarization: A method for constructing capacity-achieving codes for symmetric binary-input memoryless channels. IEEE Transactions on Information Theory **55** (7), 3051–3073 (2009). DOI 10.1109/TIT.2009.2021379
4. Atwood, G.: Current and emerging memory technology landscape. In: 2011 Flash Memory Summit. Santa Clara, CA (2011)
5. Cheng, M.H.: Generalised berlekamp-massey algorithm. Communications, IEE Proceedings-**149**(4), 207–210 (2002). DOI 10.1049/ip-com:20020413
6. Chien, R.: Cyclic decoding procedures for bose- chaudhuri-hocquenghem codes. Information Theory, IEEE Transactions on **10**(4), 357–363 (1964). DOI 10.1109/TIT.1964.1053699
7. Emre, Y., et al.: Enhancing the reliability of STT-RAM through circuit and system level techniques. In: Signal Processing Systems (SiPS), 2012 IEEE Workshop on, pp. 125–130 (2012). DOI 10.1109/SiPS.2012.11
8. Ferrari, M., Amato, P., Bellini, S., Laurent, C., Sforzin, M., Tomasoni, A.: Embedded ECC solutions for emerging memories (PCMs). In: DATE2016. Technische Universität Kaiserslautern (2016). URL urn:nbn:de:hbz:386-kluedo-43206
9. Graham, R.L., Knuth, D.E., Patashnik, O.: Concrete Mathematics: A Foundation for Computer Science, 2nd edn. Addison-Wesley Longman Publishing Co., Inc., Boston, MA, USA (1994)

10. Hamming, R.: Error detecting and error correcting codes. Bell System Technical Journal, The **29**(2), 147–160 (1950). DOI 10.1002/j.1538-7305.1950.tb00463.x
11. Henkel, W.: Another description of the berlekamp-massey algorithm. Communications, Speech and Vision, IEE Proceedings I **136**(3), 197–200 (1989). DOI 10.1049/ip-i-2.1989. 0028
12. JEDEC: Low power double data rate 4 (LPDDR4) (2014). URL https://www.jedec.org/ JESD209-4
13. Ji, B., et al.: In-Line-Test of Variability and Bit-Error-Rate of HfOx-Based Resistive Memory. In: Memory Workshop (IMW), 2015 IEEE International, pp. 1–4 (2015). DOI 10.1109/IMW. 2015.7150290
14. Kraft, C.: Closed solution of Berlekamp's algorithm for fast decoding of BCH codes. IEEE Trans. Info. Theory **39**(12), 1721–1725 (1991)
15. Lee, H.: High-speed VLSI architecture for parallel Reed-Solomon decoder. Very Large Scale Integration (VLSI) Systems, IEEE Transactions on **11**(2), 288–294 (2003). DOI 10.1109/ TVLSI.2003.810782
16. Lin, S., Costello, D.J.: Error Control Coding, Second Edition. Prentice-Hall, Inc., Upper Saddle River, NJ, USA (2004)
17. Lint, J.H.v.: Introduction to Coding Theory. Springer-Verlag New York, Inc., Secaucus, NJ, USA (1982)
18. Liu, T., et al.: A 130.7mm2 2-layer 32 Gb ReRAM memory device in 24 nm technology. In: Solid-State Circuits Conference Digest of Technical Papers (ISSCC), 2013 IEEE International, pp. 210–211. S. Francisco, CA (2013). DOI 10.1109/ISSCC.2013.6487703
19. MacWilliams, F., Sloane, N.: The Theory of Error Correction Codes. North-Holland, Amsterdam (1977)
20. Mann, H.B.: On the number of information symbols in bose-chaudhuri codes. Information and Control **5**(2), 153–162 (1962). DOI http://dx.doi.org/10.1016/S0019-9958(62)90298-X. URL http://www.sciencedirect.com/science/article/pii/S001999586290298X
21. Mastrovito, E.: VLSI designs for multiplication over Finite Fields GF(2^m). Lecture Notes in Computer Science, Springer-Verlag **357**, 297–309 (1989)
22. McEliece, R.J.: Finite fields for computer scientists and engineers. The Kluwer international series in engineering and computer science. Kluwer Academic Publishers, Boston (1987). URL http://opac.inria.fr/record=b1086077. Réimpressions : 1995 (second printing), 2003 (Sixth Printing)
23. Moon, T.K.: Error Correction Coding: Mathematical Methods and Algorithms. Wiley-Interscience (2005)
24. Mott, N.F.: Metal-Insulator Transitions, Second Edition. Taylor & Francis, Inc., 4 John St, London, GB (1990)
25. Oh, T.Y., et al.: A 3.2 Gb/s/pin 8 Gb 1.0 V LPDDR4 SDRAM with integrated ECC engine for sub-1 V DRAM core operation. In: Solid-State Circuits Conference Digest of Technical Papers (ISSCC), 2014 IEEE International, pp. 430–431 (2014). DOI 10.1109/ISSCC.2014. 6757500
26. Otsuka, W., et al.: A 4 Mb conductive-bridge resistive memory with 2.3 GB/s read-throughput and 216 MB/s program-throughput. In: 2011 IEEE International Solid-State Circuits Conference Digest of Technical Papers (ISSCC), pp. 210–211. S. Francisco, CA (2011)
27. Ryan, W.E., Lin, S.: Channel codes: classical and modern. Cambridge Univ. Press, Leiden (2009)
28. Strukov, D.: The area and latency tradeoffs of binary bit-parallel BCH decoders for prospective nanoelectronic memories. In: Proc. of 40th Asilomar Conference on Signals, Systems and Computers (ACSSC), pp. 1183–1187. Pacific Grove, CA (2006)
29. Tsuchida, K., et al.: A 64 Mb MRAM with clamped-reference and adequate-reference schemes. In: 2010 IEEE International Solid-State Circuits Conference Digest of Technical Papers (ISSCC), pp. 258–259. S. Francisco, CA (2010). DOI 10.1109/ISSCC.2010.5433948

30. Udayakumar, K., et al.: Low-power ferroelectric random access memory embedded in 180 nm analog friendly CMOS technology. In: 2013 5th IEEE International Memory Workshop (IMW), pp. 128–131. Monterey, CA (2013). DOI 10.1109/IMW.2013.6582115

31. Villa, C., et al.: A 45 nm 1 Gb 1.8 V phase-change memory. In: 2010 IEEE International Solid-State Circuits Conference Digest of Technical Papers (ISSCC), pp. 270–271. S. Francisco, CA (2010). DOI 10.1109/ISSCC.2010.5433916

Emerging Memories in Radiation-Hard Design

Roberto Gastaldi

1 Introduction

In many applications, electronic equipment must operate in an environment exposed to heavy radiation. These environments include those of: satellites in a terrestrial orbit used for telecommunication or military tasks and spacecraft on scientific missions outside our solar system in deep space; civilian and military aircraft; and equipment for measurement and control during experiments in high-energy physics and near nuclear power plants, industrial accelerators, and sources and in medical applications like radiology and radiotherapy. However, even electronic equipment used for common consumer applications can be affected by low-dose radiation, e.g., due to natural radioactivity in materials, high energy cosmic rays (especially in avionic applications), natural contaminants, and X-Rays coming from scanners in airports and controlled zones. In particular, commercial integrated circuits (ICs) that are the components of electronic systems can have a high sensitivity to radiation, so that their reliability is greatly compromised. IC technological evolution, which tends to yet denser integration, has generated two opposite phenomena: on one hand, devices became more tolerant to cumulative effects of radiation, while, on the other, they are more sensitive to soft errors generated by single events, because the dimensions of critical nodes have been reduced.

For many years, the design community has tried to answer the question about how to protect integrated circuits against radiation effects; in this regard, we can distinguish different approaches. The easiest conceptually is to shield the electronic systems with materials that stop radiation; this concept can be scaled down at component level using packages particularly designed to shield the internal die. Although this method is not invasive to the system and its components (ICs), it has

R. Gastaldi (✉)
Redcat Devices s.r.l, Milano, Italy
e-mail: r.gastaldi@redcatdevices.it

© Springer International Publishing AG 2017
R. Gastaldi and G. Campardo (eds.), *In Search of the Next Memory*,
DOI 10.1007/978-3-319-47724-4_9

a big limitation due to the weight of the shielding. If high protection is needed, thick metal shields must be used, whose weight cannot be tolerated in avionics and space applications, plus it is not easy to shield all harmful radiation this way. The second method is to operate at component level, making modifications to the silicon technology and/or to design to achieve immunity to radiation. If special technological steps are introduced to increase robustness, we speak about RHBP: radiation hardening by process. Otherwise, in cases where we rely on special design methodology to meet the target, we talk about RHBD (radiation hardening by design).

2 Radiation Environments

High-energy particles from galactic cosmic rays interact with the outer terrestrial atmosphere, losing energy and giving rise to a shower of secondary particles that strike the Earth's surface, in particular, neutrons and muons. Neutrons are the most important concern at sea level. Another source of radiation on the Earth's surface are the natural contaminants present in materials used for ICs' packages.

In the interplanetary space, instead, radiation is composed of electrons and protons trapped in the magnetic field of planets, high-energy particles coming from the Sun (protons and heavy ions), and cosmic rays (protons and heavy ions).

The interaction between external radiation (photons like X-rays and γ-rays, charged particles like protons, electrons, and heavy ions, or neutral particles) and a semiconductor cause two main phenomena: ionization and displacement. The first phenomenon creates an electron-hole pair (HEP), the second effect is an energy transfer from the incoming radiation and the reticule, causing modification to the reticule itself.

3 Effects of Radiation on Semiconductors

When radiation interacts with the semiconductor material, energy transferred to the reticle leads to the generation of an electron-hole pair by ionization: an electron acquires enough energy to move to conduction band as a free electron, leaving a hole in the valence band. If this event takes place in a region where an electric field is present, pair is separated and carriers move under the action of the electric field giving a transient current, before eventually recombine, or remain trapped, usually into an insulator, or they are collected from an electrode. In Fig. 1, the ionization process and resulting charge kinetics are illustrated.

Energy lost by an incident particle during its impact on material is measured by Linear Energy Transfer (LET). The LET depends on atomic number and energy of the striking particles and on the target material, and it is defined as

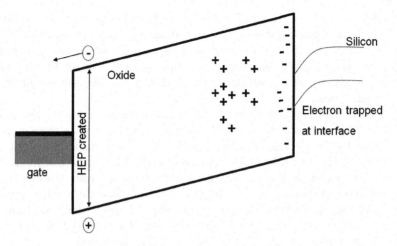

Fig. 1 Charge generated by radiation is trapped inside the oxide

$$LET = (dE/dx)1/\rho \quad (Me\ cm^2/mg) \tag{1}$$

where ρ is the density of the target material and dE/dx indicates the average energy transferred into the target material per length unit along the particle trajectory. Another possibility is that a particle striking the silicon can displace an atom from its original position, leaving a defect in the reticle that can be seen as the creation of localized energy levels in the band-gap altering electrical properties of semiconductor. Part of the charge generated during ionization phenomena described above remains trapped in insulator layers [1–3] of a silicon device or at the interface between insulator and semiconductor where free carriers have a very low mobility and are easily captured by traps or interface states. In particular holes which have mobility many orders of magnitude lower than electrons can easily be trapped and act as a fixed positive charge both in gate oxides and in field oxides. Damaging effects caused by this trapped charge are cumulative and become important after a long exposure to radiation, they are measured with the Total Ionizing Dose (TID) which is the total amount of energy imparted by ionizing radiation to a unit mass of absorbing material and it is measured in Rads (1 rad = 0.01 J/kg).

In CMOS integrated circuits, the region most sensitive to cumulative effects is the gate oxide. The trapped holes (positive charge) introduce a negative shift in threshold voltage ΔV_{th}, given by:

$$\Delta V_{th} = -q/C_{ox}\Delta N_t = -(q/\varepsilon_{ox})t_{ox}\Delta N_t \tag{2}$$

where q is the elementary charge, $C_{ox} = \varepsilon_{ox}/t_{ox}$ is the oxide capacitance per unit area, N_t is the density of trapped holes into the oxide, ε_{ox} is the dielectric constant of the oxide, and t_{ox} is the oxide thickness. Effects of negative threshold shift [4] cause an increase of sub-threshold current of N-MOS transistors and decrease their controllability through the gate. However, at the first degree of approximation, trapped charge is proportional to t_{ox}, and consequently ΔV_{th} is proportional to t_{ox}^2, and for thin or very thin oxides (e.g., for thicknesses thinner than approximately 3 nm), threshold shift becomes negligible. This means that threshold shift due to charge trapped into the gate oxide is more important for older technologies using thicker gate oxides.

A second problem is coming from positive charge trapped in the Shallow Trench Isolation (STI) regions at the transition between a field-thick oxide and a gate-thin oxide (see Fig. 2). The region on the side of an STI can be modeled as a parasitic transistor in parallel with the MOS transistor channel. They are normally turned off because, due to the thick oxide their threshold voltage is higher, however due to positive charge trapped causing a negative shift of threshold voltage, parasitic transistors could turn on thus creating a parasitic path between drain and source, in parallel with the MOS-transistor channel once again increasing leakage current. Unfortunately this second effect is not improving too much with technology scale-down because field oxide is thick enough to cause non-negligible threshold voltage shift.

Also Displacement Damage Dose (DDD) has an effect of MOS electrical characteristics: lattice defects created at the Si-SiO$_2$ interface by the displacement introduce energy states in the band-gap [5, 6], which may trap channel carriers. The voltage threshold shift due to the charge trapped into interface states is

$$\Delta V_{IT} = -Q_{IT}/C_{ox} \tag{3}$$

where Q_{IT} is the trapped charge at the interface, which depends on device biasing. Moreover, trap states due to lattice defects facilitate decrease carriers mobility:

$$\mu = \mu_0/(1 + \alpha\Delta N_{IT}) \tag{4}$$

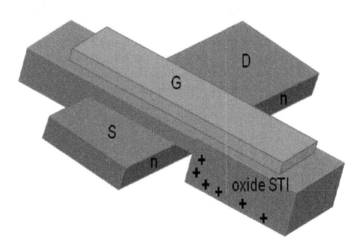

Fig. 2 Positive charge trapped in STI oxide

where μ_0 is the pre-irradiated mobility; α is a parameter dependent on the chosen technology; and ΔN_{IT} is the number of charges trapped at interface.

As a conclusion in an NMOS transistor, TID causes a leakage current when the transistor is turned-off and an alteration of transconductance characteristic with a decrease of gain β defined as:

$$\beta = \mu C_{ox} \tag{5}$$

A comparison of I_d/V_g characteristic before and after radiation exposure is shown in Fig. 3 where it can be seen the big increase of sub-threshold current after irradiation.

Ultimately these effects cause a failure of the whole system, but, even before this point, higher leakage current can cause power consumption get out of specification, particularly in stand-by mode.

In a PMOS transistor, TID causes an increase of the threshold voltage, but usually this effect is less dangerous for device functionality.

So far we have seen the effects due to a cumulative absorption of radiation that are more and more severe increasing the exposition time. There is another class of effects [7, 8] called single event effects (SEE) that are generated by a single strike of a high-energy particle at a point of a semiconductor, this class of effects can cause transient failures called SEU (single-event upset) to electronic devices or result in a permanent failure or damage, then we speak of SEL (single-event latch-up) or SEGR (single-event gate rupture).

Single-event effects are due to charge generation in a reverse-biased p-n junction in a CMOS IC. The junction may be part of a MOS transistor (drain-body or source-body), or may be a well-substrate junction. This mechanism is illustrated in

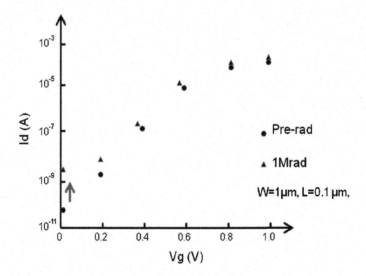

Fig. 3 Id/Vg shift of a NMOS transistor after irradiation

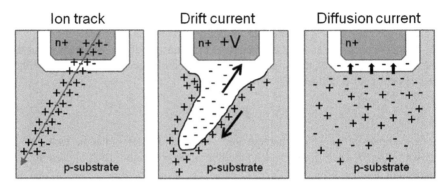

Fig. 4 Hole-Electron pairs generation in the substrate as a result of the impact with a high-energy particle. Part of the charge can be collected by biased junctions causing errors and malfunctions

Fig. 4 where a high-energy ion strikes a p-n junction generating a cloud of hole-electron pairs. The electric field in the reverse-biased p-n junction separates electrons and holes. The generated carriers are collected by neighboring electrodes, thus imparting a parasitic current with a peak due to carrier drift, followed by a tail due to carrier diffusion. N-junctions at a positive voltage collect the generated electrons, and, if the collected charge is higher than a critical charge, generates a glitch, temporarily changing the logic state of the collecting node: this is called a single-event transient (SET). Most of the logic nodes allow the transient to be recovered in a few nanoseconds as shown in Fig. 5, and, in pure combinatorial logic, the impact of a SET can be limited; however, they may propagate to adjacent

Fig. 5 Transient current generated by radiation can be recovered in few nanoseconds

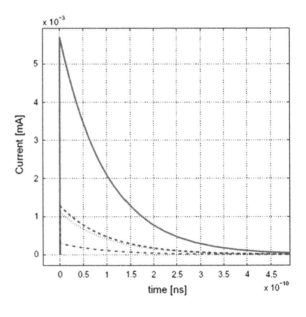

nodes where the effect of other SETs can be added; if parts of the circuitry containing sequential logic are affected (latches, registers, or memories) [9], a permanent change of logic state in some nodes or data toggling in the memory can occur (see Fig. 6) and a single-event upset (SEU) takes place.

In this case, recovery is not possible because the transient has triggered a permanent logic change in some nodes. At some point, this logical error can occur after a number of logic operations involving different circuits, and it may be difficult to identify the point at which the original malfunction took place.

If a SEU affects two or more memory cells, a multiple bit upset (MBU) occurs. A SEU in the control logic may lead to a single-event functional interruption (SEFI).

Unfortunately, sometimes the sum of SET's can trigger physical damage in semiconductors, and this happens more frequently in power circuits where high currents or voltages are involved. In a structure as the one schematized in Fig. 7, the charge collected in the substrate after a SEE can trigger a positive gain loop made of parasitic bipolar transistors, resulting in a single-event latch-up (SEL), and high current flowing in the circuit during latch-up may lead to complete destruction of the IC [10].

Other destructive SEEs are the single-event burnout (SEB), which occurs in high-voltage devices when an avalanche multiplication mechanism is triggered by a parasitic charge in a p-n junction reverse biased and a single-event gate rupture (SEGR), where the displacement effect can result in an oxide-gate rupture. SEB and

Fig. 6 A SET can propagate and generate a SEU in another circuit

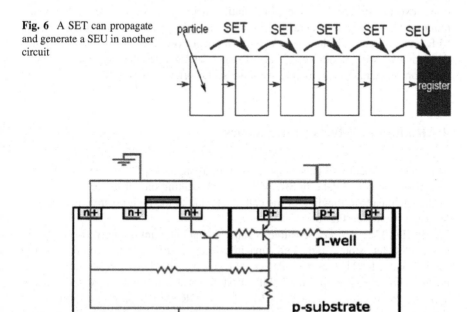

Fig. 7 Parasitic bipolar components in CMOS technology

Table 1 Summary of damaging events in semiconductor circuits

SEU	Single-Event Upset	Change of bit in memory element	Memory, latch
SET	Single-Event Transient	Temporary variation of a bit or of a voltage value	Analog and digital
MBU	Multiple-Bit Upset	Change of several bits coming from a SET/SEU	Memory, latch
SEFI	Single-Event Failure Interrupt	Corruption of data path in state machines	Flash memory, microprocessor, FPGA
SED	Single-Event Disturb	Temporary corruption of a bit coming from a SET	Combinational logic
SEL	Single-Event Latchup	High-current conditions leading to hard error	CMOS, BiCMOS
SEGR	Single-Event Gate Rupture	Rupture of the gate coming from high-electric field	Power MOSFET, NV memory, Deep VLSI
SEB	Single-Event Burnout	Burnout coming from high-current conditions	BJT, n-channel Power MOSFET
SHE	Single-Event Hard Error	Unalterable change of a bit	Memory (NV)
SESB	Single-Event Snapback	High-current conditions	Power MOSFET, SOI

SEGR occur in power MOS transistors [11,12] and are minor problems for memories, except perhaps in some non-volatile memory in which the high voltage required for writing is generated on-chip using charge pumps. Sensitivity versus SEE is measured with the cross section (in square centimeters), which constitutes the sensitive area of device. Finally Table 1 summarizes the possible effects of radiation on semiconductors that we have so far briefly discussed.

4 Radiation Effects in Memories

SEE effects in memories have been studied for a long time [13, 19], because of their big impact on stored data. In SRAM, a particle striking one of the feedback nodes of the latch leads to toggling of the cell's data due to the high-positive feedback of the latch itself (see Fig. 8).

Also DRAMs are very sensitive because a small amount of charge is enough to destroy the data stored in the cell capacitor. Even if SEE are much more critical in SRAM and DRAM, they can also be affected by cumulative effects due to total dosage (TID): NMOS threshold shift in 6T-RAM can degrade the stability factor and increase leakage current [14, 16], while in DRAM degrading performance of the cell's selector can reduce retention time. The periphery can be even more affected because larger transistors are present.

Fig. 8 Electrical circuit of a typical SRAM cell. Radiation particles striking critical nodes, as those indicted by the *arrows*, can cause flipping of memory-cell state inducing a bit failure in the memory array

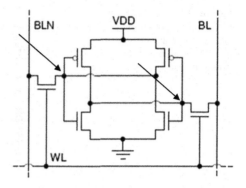

For FLASH memories TID represents an important issue due to the fact that the storage mechanism relies on a charge accumulated in a floating gate surrounded by insulator. The technology's scaling-down roadmap reduces the charge stored in the cell to hundreds or tens of electrons, thus increasing the sensitivity to radiation effects.

In Fig. 9, a summary of these effects is presented, for a cell that is supposed to store a charge Q_{st} in the floating gate. The first effect is the creation of electron-hole pairs in the floating gate and control- gate oxide. Electrons in conduction band are swept out from the electric field created by the negative charge stored in the floating gate, while holes are collected in the floating gate, compensating the negative charge. The second one is an energy transfer from the high-energy particles striking the floating gate directly to stored electrons that are emitted over the oxide barrier (photoemission). Both these phenomena add up to cause charge loss in the cell.

Fig. 9 Effects of radiation on a flash-memory cell

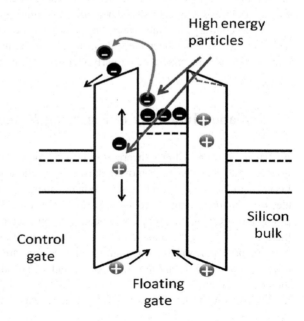

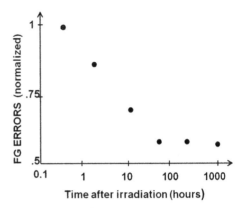

Fig. 10 Threshold shift induced by exposure to high-energy radiation can be partially recovered after a period of time

A third mechanism involves the positive charge created by ionization that can be trapped in the oxide and cause a local band bending that, on one hand, helps the charge emission by tunneling from the floating gate, and, on the other hand, increases the local electric field in the oxide, degrading reliability.

Experiments have been reported for NAND flash that confirm the impact of total dose on retention. Retention failures have been reported at dose < 50 Krad, quite independently from operating conditions.

The action of total dose on the threshold distribution in flash cells is to shift the whole threshold distribution toward the intrinsic distribution that is the one in which no charge is present in the floating gate.

Charge loss and errors consequently generated can be recovered at least for the portion due to trapped charge in the oxide, and that can be removed by annealing bake or de-trap spontaneously after a period of time (see, for example, Fig. 10).

On the contrary, in other cases the damage is permanent, probably because a defect in the oxide has been created during the strike: in this case the cell will have retention problems, experiencing early charge loss after a reprogramming.

5 Radiation Immunity of Emerging Memories

As we have seen, a big concern for flash memories is the charge stored in the floating gate that can be removed during interaction with high-energy radiation, leading to loss of stored data. When emerging memories don't rely on electrical-charge storage to retain logic data, but more on changes of bulk-material physical properties, then they could be more robust under radiation stress. Although there are no extensive data for all emerging memories under development, TID measurement performed with gamma and X-rays are reported on 90 nm PCM devices showing TID immunity up to 0.7 and >2MRad [15], while no errors after unbiased irradiation have been found until LET(Si) = 58 MeVmg^{-1}cm^2. Upsets are not expected to occur along the scaling path at least before 32-nm node. [25].

A possible degradation mechanism that has been proposed is that penetration of heavy ions inside the GST, can lead to a heating of a small cylinder of the material around the particle path due to electron-phonon coupling until melting temperature is locally reached. This can cause a local phase change and a consequent modification of overall resistance of the memory element. This failure mechanism changes SET cells to RESET [25] and should become more important reducing cell's feature size. On the other hand, an incidence of single-event latch-up (SEL) and single-event functional interrupt (SEFI) in PCM irradiated with heavy ions has been reported [15, 23] but these failures seem to be connected to layout or external circuitry rather than to the memory material.

Measurement made on oxide ReRAM shows that the switching window exhibits a degradation under protons irradiation at high fluence, but at least some of the damage can be recovered after tens of writing cycles [20, 24]. Degradation is due to displacements damage in the oxide, which randomly create vacancy defects, so then a parallel filament can be created in parallel with the pre-existing one (see Fig. 11).

The layout configuration of a ReRAM cell, including a selector NMOS transistor and similar in this aspect to MRAM and PCM (at least for one of the possible configurations), can show SEE weakness. In fact, considering a generic cell cross-section with the non-selected cell, we have the situation of Fig. 12. In this figure the gate of selector transistor is at ground, and drain n^+ region is at 1.8 V, because no current flows in the system. If a particle strikes the drain region of selector, a glitch to GND can be produced and enough bias voltage would appear across the cell to cause a spurious write.

Another criticality linked to the selector is the possibility of a threshold shift and degradation of the transconductance that would impact the leakage current through the unselected cells and could eventually lead to bit failure. The same kind of degradation affects NMOS in peripheral circuits, and we have seen that the reason for device failure can originate more from a malfunction in periphery circuitry than from intrinsic weakness of the memory cell, so in the following we will cover, in a little more detail, the ways to protect circuitry from radiation effects.

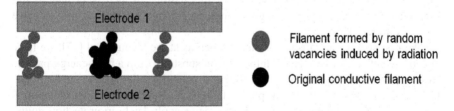

Electrode 1

Electrode 2

● Filament formed by random vacancies induced by radiation

● Original conductive filament

Fig. 11 Mechanism of degradation caused by interaction with high-energy protons in a ReRAM cell

Fig. 12 Problem of disturb
in a ReRAM cell

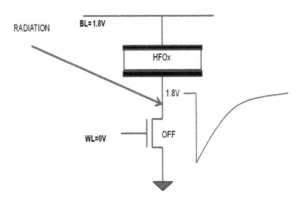

6   Design of Radiation-Hardened Memories

In the previous discussion about memory robustness to radiation stress, it is clear
that, beyond the specific weaknesses related to the memory cell itself, coming from
storing mechanism and materials used, an important cause of failure is represented
by the periphery circuitry that is subject to the same problems of all MOS devices
when operating in a radiation environment. In many cases, in particular in emerging
memories, these problems can cancel the advantages that come brought by an
intrinsically robust memory cell and lead to poor performance of the device.

Using a shield to protect electronic systems is very expensive due to the weight
associated with the shielding, and the inevitable trade-off with performances can be
not completely satisfactory. Another possibility comes from process modifications,
for example, the use of SOI can improve robustness to induced latch-up (SEL). In
general, in any case, adoption of special process for rad-hard devices is not sustainable by foundries, due to the relatively low volume of devices requiring high
rad-hard performance. In addition, transporting a design from one foundry to the
other is more difficult than using a full standard process. Also, technology scaling–
down, which reduces area of critical nodes and distances, makes it more and more
difficult to adequately protect devices. For this reason, an increasing interest has
been put on inventing special design techniques and layout rules to improve device
tolerance to radiation.

The first step for an effective protection begins at the layout level [17]: we have
already seen that charge injected into a silicon substrate from a high-energy particle
striking a semiconductor device can trigger a potentially destructive latch-up. We
have also seen that the prevalent part of failures in experiments under radiation is
due to SEL. So the natural countermeasure is to apply latch-up protection guide-
lines whenever possible, making guard-rings and reinforcing contacts to Vdd and
gnd to avoid that some junctions be biased in such a way to make latch-up easier.

An example of this kind of treatment is shown in Fig. 13.

Here two n^+ and p^+ guard-rings around a p-well and an n-well are interposed
between a p-channel and an n- channel of an MOS to prevent latch-up triggering.

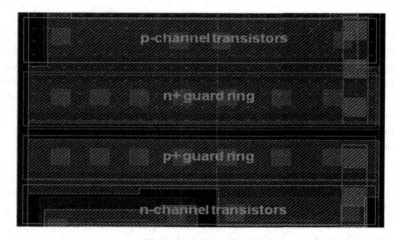

Fig. 13 Layout of guard-rings between p-channel and n-channel transistors to prevent latch-up

Guard-rings must be connected to Vdd and gnd respectively through a metal line above them and the contacts chain to guarantee that all the ring is at the same potential. In addition, whenever possible, n-channel and p-channel MOSs should be grouped and positioned as far from each other as much as possible. Moreover, it is possible to use guard-rings around transistors of the same type but biased at different voltages to reduce the leakage induced by positive charge trapped in the STI oxide.

We have also seen in the previous discussion that a big problem is the charge accumulated at the edge of STI, which creates a leakage path bypassing the transistor channel. This is a long-term effect depending on the total dose absorbed (TID) and produces a degradation of the transistor characteristics that eventually makes it impossible to switch off the device, leading to component functional failure. To counteract efficiently this issue, it is possible to adopt the so-called edge-less transistor (ELT) that is a way to layout the transistor without an edge on the field oxide. In this transistor the drain region is completely surrounded by a source region and the anular gat defines the channel region, so that no field edge crosses the gate and a parallel leakage path is no longer possible.

Generally, it is convenient in ELT transistor to minimize the drain area that is the most sensitive node to SEE, which means to put the drain terminal on the internal side.

Although ELT is an effective way to increase TID robustness of MOS ICs, the price to pay in terms of area is very high. It should be noted that ELT is mandatory only for NMOS because trapped charge is positive, PMOS are not affected, and this contributes to mitigate the area drawback. But still it remains a problem for more dense and performance demanding ICs, not only for the mentioned area occupation, but also due to the increase of parasitic capacitances and gate capacitance that prevent very high-speed operation.

Looking at memory ICs that are built around the array of cells, the use of ELT in decoding circuitry puts a limit on the reduction of row and column pitch, in particular for flash memories where the last stage of decoding must sustain the high voltage required for programming. Emerging technologies are in a better situation because they require much less voltage than flash during write. On the other hand, if ELT is used, the memory cell itself obviously cannot be made very small. SRAM cells using extensive ELT proved to be immune up to TID = 15Mrad and SEU LET immune up to $5MeVmg/cm^2$ while avoiding ELT in all the transistors TID immunity has been reduced to <200Krad.

At circuit level, a designer can use dedicated circuit configurations to lower sensitivity to SEE. Low capacitive nodes are more sensitive to the transients caused by particles strike (SET), so full C-MOS logic should be used as much as possible, and the number of transistors not directly connected to supplies should be minimized. Also, very sensitive nodes may be designed to have a higher parasitic capacitance in such a way to increase critical charge.

An important aid for designers comes from dedicated software that can simulate the effect of charge injection in a node of the network. It becomes possible to evaluate the critical charge for nodes that are judged particularly critical and understand how the generated disturbance propagates throughout the whole network and its effect on the global operation [18].

Mitigation of SEFI requires that feedback loops are avoided as much as possible, but this cannot be done easily when sequential logic is involved, so a possibility is to try to delay the loop to give the transient the time to finish [19]: unfortunately, the time constants involved may be a serious problem for high speed logic.

A sort of hardware redundancy is represented from voter circuits as depicted in Fig. 14.

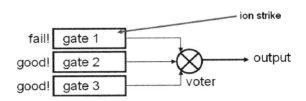

Fig. 14 Concept schematic of voter circuitry

The concept is to replicate the same gate or logic circuit an odd number of times (three in the figure) and take as correct the output indicated by the majority of the gates, considering that the probability that a particle strikes the majority of the gates at the same time is low enough.

As a matter of fact, all these techniques pay a high price in silicon area and operation speed, so a trade-off is always needed to target of robustness and electrical performance.

Concerning memories, there are additional considerations to be made for chip architecture: it has been found that the charge pump used to generate the on-board high voltage required for flash-cell programming is particularly weak under radiation stress, even at relatively low doses of irradiation, causing the memory device to fail program. This is an issue favoring emerging memories that don't require high voltage for programming.

We have already seen that the cells inside the array can fail due to a SEE effect. This failure can be seen from the outside as a single bit failure or a pair failure if the neighbor bit is affected.

An efficient way to survive with this kind of error is to provide an error detection and correction (EDAC) engine on board; this is quite common not only for rad-hard memories, but also in consumer products, such as large-capacity NANDs, and it is mandatory in almost all emerging technologies to achieve an acceptable low-bit error rate.

The overhead of EDAC-blocks for circuit complexity and access- time delay depends on the number of fail bits that can be corrected in a word, and, unfortunately, memories in a radiation environment are sensitive to failures that affect many bits or even the whole word. For example, an SEU occurring in the array row or column decoding will lead to an incorrect reading or programming of the whole row or column, if they are selected during the upset. In case of hard damage, the failure of the selected element will be permanent. It is then reasonable to prefer array architectures that mitigate the risk of failure of a whole word or column [22].

Significant improvement can be obtained with an array partitioning, by providing each memory array storing a single bit of the word with separate bit-line and word-line decoders, to avoid MBUs. An example of such an array organization in a 512K SRAM is shown in Fig. 15. Even if many cells are affected by a particle strike, the resulting error will be only on a single bit of the word, and this is true also if one of decoders is hit.

On the other side an issue of this solution is the large area occupation that limits the maximum memory size practically achievable.

In addition it requires a careful design to balance different signal paths at byte level and to avoid cross-coupling noise among the tiles. Non-volatile memories, as we have seen, if they are affected by a single event or by cumulative effect, can degrade the information stored. For example, flash can experience charge loss, and

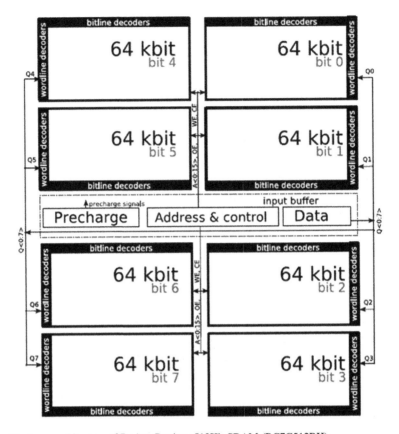

Fig. 15 Array architecture of Redcat Devices 512Kb SRAM (RC7C512RH)

PCM and other emerging memories can change their resistance or can go through a write disturb as described above for ReRAM. In this case, the signal available for reading the memory is reduced and the result can be a reading failure or an unstable reading that can give a correct result or a failure depending on the temperature, operating conditions, and the process parameters. This kind of malfunction can be very difficult to detect in the system before it eventually evolves into a permanent fail. The concept that may be employed to mitigate read-window narrowing, is to use two cells for a single bit and compare each to the other after programming them with a complementary data, as conceptually shown in Fig. 16. Although it doubles the available reading window, this self-referencing architecture is highly area consuming and can be used for low-medium memory size.

Fig. 16 Conceptual drawing
of self-reference reading
scheme

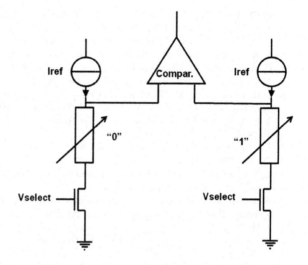

7 Conclusions

Unlike mainstream flash technology, which is affected by interactions with high-energy particles due to their mechanism being based on charge stored in a floating gate, emerging memories prove to be more robust against TID degradation. In addition, low-voltage operation required for most of emerging technologies allows the elimination of some circuits, like charge pumps that were found to be a critical point during irradiation experiments.

Nevertheless, peripheral circuitry and the cell selector device become the weak point of the system, and process and design techniques to mitigate this weakness must be adopted. In particular, radiation hardening by design has the advantage to operate on a standard process and thus to optimizing reliability and cost, but by paying a price in terms of silicon area and performance.

For these reasons, radiation hardening by design always requires a compromise between the degree of protection and the electrical performance required by the system.

Acknowledgements The author would like to thank Cristiano Calligaro, CEO of Redcat Devices s.r.l for making available some of the drawings used in this chapter.

References

1. P. J. McWhorter et Al. "Simple technique for separating the effects of interface traps and trapped-oxide charge in metal-oxide-semiconductor transistors," Appl. Phys. Lett., vol. 48, pp. 133–135, Jan. 1986.
2. N. S. Saks et Al. "Radiation effects in MOS capacitors with very thin oxides at 80 K," IEEE Trans. Nucl. Sci., vol. 31, pp. 1249–1255, Dec. 1984.

3. F. B. McLean, "A framework for understanding radiation-induced interface states in SiO2 MOS structures" IEEE Trans. Nucl. Sci., vol. 27, pp. 1651–1657, Dec. 1980.

4. M. Gaillardin et Al. "Enhanced radiation induced narrow channel effects in commercial 0.18 μm bulk technology," *IEEE Trans. Nucl. Sci.*, vol.58, pp. 2807–2815, Dec. 2011.

5. G. Baccarani et Al. "Transconductance degradation in thin-oxide MOSFET's," IEEE Trans. Electron Devices, vol. 30, pp. 1295–1304, Oct. 1983.

6. J. R. Schwank et Al. "Total ionizing dose hardness assurance issues for high dose rate environments," *IEEE Trans. Nucl. Sci.*, vol. 54, pp. 1042–1048, Aug2007.

7. T. C. May, "Soft errors in VLSI: Present and future," IEEE Trans. Comp., Hybrids, Manufact. Technol., vol. 2, pp. 377–387, Dec. 1979.

8. J. L. Andrews et Al. "Single event error immune CMOS RAM," IEEE Trans. Nucl. Sci., vol. 29, pp. 2040–2043, Dec. 1982.

9. L. T. Clark et Al. "Optimizing radiation hard by design SRAM cells," IEEE Trans. Nucl. Sci., vol. 54, pp. 2028–2036, Dec. 2007.

10. Johnston, "The influence of VLSI technology evolution on radiation induced latchup in space systems," IEEE Trans. Nucl. Sci., vol. 43, pp. 505–521, Apr. 1996.

11. J. H. Hohl et Al. "Analytical model for single event burnout of power MOSFETs," IEEE Trans. Nucl. Sci., vol. 34, pp. 1275–1280, Dec. 1987.

12. F. Wheatley et Al. "Single-event gate rupture in vertical power MOSFETs; an original empirical expression," IEEE Trans. Nucl. Sci., vol. 41, pp. 2152–2159, Dec. 1994.

13. G. Anelliet Al. "Radiation tolerant VLSI circuits in standard deep submicron CMOS technologies for the LHC experiments: practical design aspects,"IEEE Trans.Nucl.Sci., vol. 46, pp. 1690–1696, Dec. 1999.

14. C.Calligaro et Al. "A multi-megarad, radiation hardened by design 512 kbit SRAM in CMOS technology," in Proc. IEEE Int. Conf. on Microelectronics (ICM), Cairo, Egypt, Dec. 2010, pp. 375–378.

15. A. Paccagnella et Al. "studies of radiation effects on new generations of non-volatile memories" ESA-ESTEC, The Netherlands, June 2013.

16. M. Benigni et Al. "Design of rad-hard SRAM cells: A comparative study," in Proc. IEEE Int. Conf. on Microelectronics (MIEL), Niš, Serbia, May 2010, pp. 279–282.

17. Stabile et Al. "A radiation hardened 512 kbit SRAM in 180 nm CMOStechnology" in Proc. Int. Conf. on Electronics, Circuits and Systems (ICECS), Hammamet, Tunisia, Dec. 2009, pp. 655–658.

18. E. Do et Al. "Layout-oriented simulation of non-destructive single event effects in CMOS IC blocks," in Proc. Radecs2009.

19. "Radiation effects in semiconductors" edited by K. Iniewski, CRC press, 2010.

20. A.Fantini "Radiation hardness of memristive systems" Workshop on memristive systems for space applications, ESTEC, Apr. 2015.

21. G.Anelli et Al. "Radiation tolerant VLSI circuits in standard deep submicron CMOS technology for the LHC experiments: practical design aspects" IEEE Trans. Nucl. Sci., vol.46 pp. 1690–1696, Dec.1999.

22. C. Calligaro " design of integrated circuits for space applications" STAR-CAS2015, Pavia, June 2015.

23. WU Liang-Cai et Al. "Total Dose Radiation Tolerance of Phase Change Memory Cell with GeSbTe Alloy" Chin.Phys. Lett. Vol.23, n.9(2006),2557.

24. S.L.Weeden-Wright et Al. "TID and Displacement Damage Resilience of 1T1R HfO2/Hf resitive memories" IEEE Trans. Nucl. Sci. vol.61, n,6 Dec.2014, pp. 2972–2978.

25. S. Gerardin et Al. "Single Event Effects in 90-nm Phase Change Memories" IEEE Trans. Nucl. Sci, vol.58, n.6, Dec.2011.

Index

© Springer International Publishing AG 2017   247
R. Gastaldi and G. Campardo (eds.), *In Search of the Next Memory*,
DOI 10.1007/978-3-319-47724-4

Printed in the United States
By Bookmasters